Reflective Geometry

Activities with the GeoReflector™ Mirror

Grades 5-8

by

E. H. Giesecke

Illustrations by: Tirza Ernst
Design by: Ellen Hart/Concepts Plus, Inc.

ISBN: 1-56911-917-1

Printed in China.

Table of Contents

Introduction To *GeoReflector* Activities

Motion and Image Manipulation

Symmetry and Congruence

Geometry and the *GeoReflector*

Selected Answers and Solutions

Introduction

This book is written for teachers and parents who wish to use Learning Resources' *GeoReflector*™ to enhance a student's understanding of basic geometry and geometric concepts. The *GeoReflector* allows for viewing images reflected by a semi-transparent plastic window. The *GeoReflector* provides opportunities for the intuitive exploration of geometry using a hands-on tool. The activities in this book can be used with other similar plastic reflective mirrors.

The blackline masters are suitable for students in grades 5 though 8. The masters present a variety of formal and informal explorations into basic geometric concepts, and can be used as springboards for explorations by individuals, partners, small groups, or an entire class.

The activities in this book address the following standards from NCTM's *Curriculum and Evaluation Standards for School Mathematics* (1989):

- Mathematics as Problem Solving
- Mathematics as Communication
- Mathematics as Reasoning
- Mathematical Connections
- Geometry and Spatial Sense

As students engage in the activities and complete constructions, they will use critical thinking skills including observation, identification, classification, prediction, estimation, recognizing cause and effect, evaluation, revision, comparison, and organizing and recording data.

Along with each activity, most blackline masters offer an additional challenging question. These additional questions allow students opportunities to extend their learning and practice their communication skills. The extension activities also allow students the chance to use their own experience and observation to formulate formal geometric concepts in their own words. Detailed Teaching Strategies and Assessment Guidelines offer guidance and suggestions for activity management.

Introduction To GeoReflector Activities

The *GeoReflector* performs best when it is used on a flat surface. We recommend the blackline masters be copied onto white paper. Viewing images in the *GeoReflector* is easiest when the *GeoReflector* is clean. (A clean, damp, soapy cloth will do the trick.) The clearest, most accurate images are observed when the viewer is at eye level with the *GeoReflector.*

Activities in this book are arranged in order of increasing difficulty. Often activities build upon understanding gained through the completion of earlier activities. There has been an attempt to limit the number of sophisticated geometric terms needed by students to understand and complete these activities. In some cases, definitions are provided within the activities. In other cases, it is left to the teacher's discretion to present the terms to students.

The first few exercises concentrate on mapping, and introduce the terms necessary for proper use of the *GeoReflector.* The activities are designed to introduce students to the *GeoReflector*, to acquaint them with the images it produces, and to help them acquire the skill needed to draw the images formed by the *GeoReflector*. These forays also lay the groundwork for understanding image turns, slides, and flips, as well as mapping, symmetry, and congruence.

Pages 7, 8, and 9 offer basic instructions for using the *GeoReflector,* and the vocabulary needed for completing future activities. Guidance is provided for recognizing the relationship between object and image. The activities informally address symmetry and congruence, concepts which will be revisited in more detail on later pages.

Pages 10, 11 simply structure explorations with the *GeoReflector.* By asking students to place an image at a specific location, students' attention is directed at the cause-and-effect relationship of object and image movement. Following are answers to the exercises on these pages.

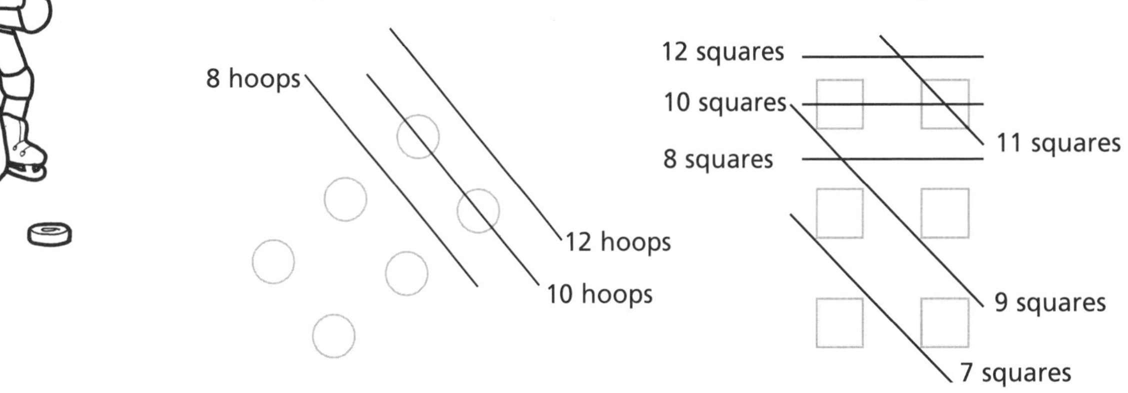

Introduction To GeoReflector Activities

Page 12 allows students to practice combining objects and images to form an entire item or secret message. Invite students to work together to design secret messages using the *GeoReflector.*

Pages 13 and 14 allow students to practice drawing images formed by the *GeoReflector.* Be sure that as students work with the arrow on page 14, they understand why the ←——←—— combination is not possible. (The reflected arrow must point in the opposite direction of the object arrow.) You might encourage students to write an explanation and then have a partner test the explanation using the *GeoReflector.*

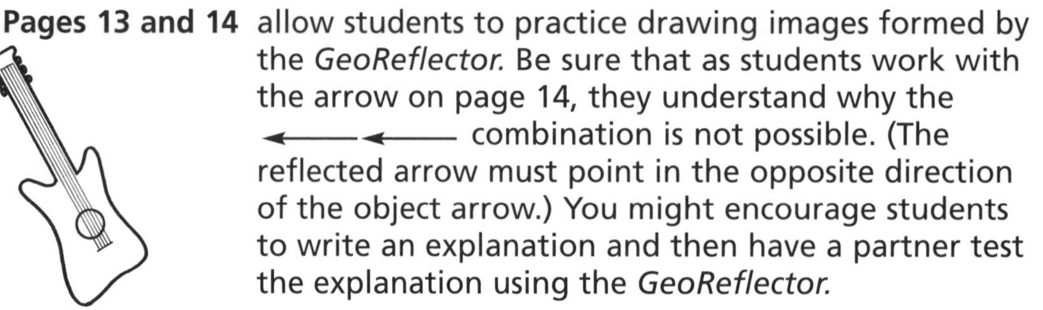

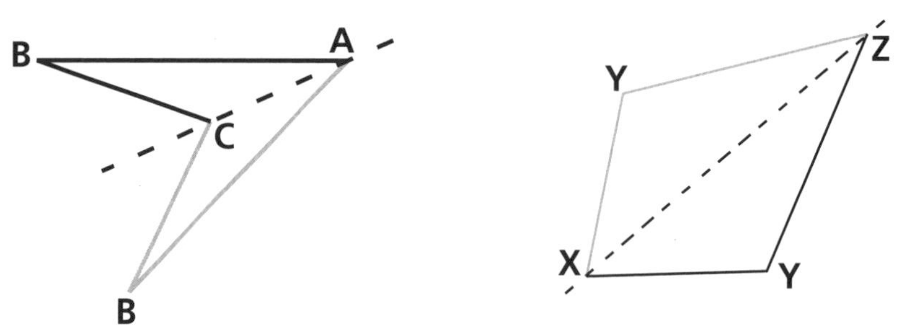

Pages 15 and 16 allow for further exploration of the object-image relationship. Students look through the *GeoReflector* and "trace" the image they see. The relationship becomes more concrete as students use an image they have generated as an object to form yet another image.

Pages 17 and 18 provide a prelude to flips and rotations by allowing students to explore what happens as an image is reflected repeatedly. You may want to relate the left- and right-handedness of the images in the *GeoReflector* to the left- and right-reversal of objects seen in an ordinary mirror.

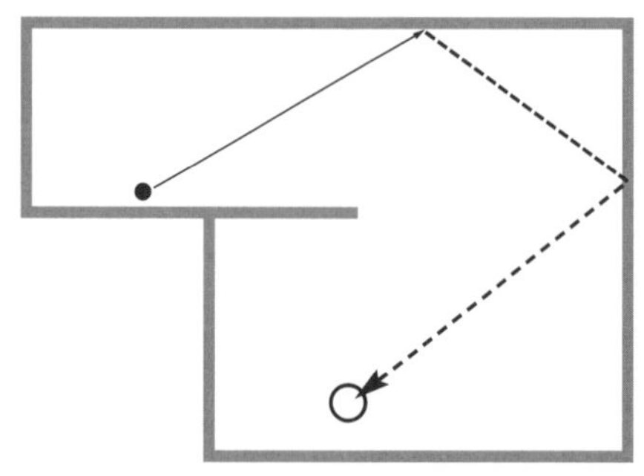

GeoReflector Anatomy

Look closely at the *GeoReflector.* Notice that it has two curved sides and two flat sides. Find the thinner edge of the *GeoReflector.* This edge is made so you can rest a pencil against it as you draw a straight line.

Place your *GeoReflector,* a ruler, a button, and a square counter so they match the drawing. Is the button in front of or behind the *GeoReflector?*

Always work with the curved sides facing you unless the directions say otherwise.

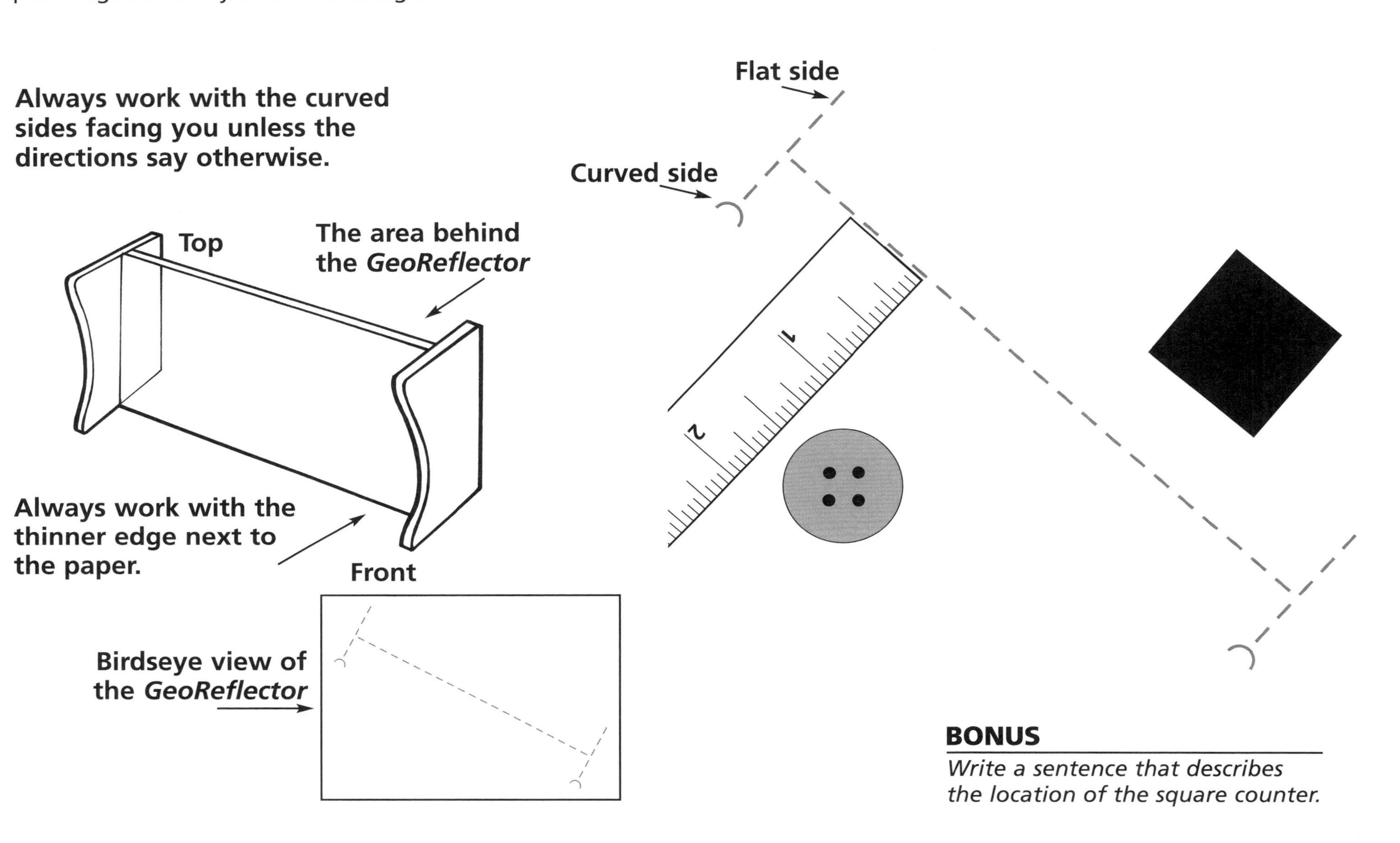

BONUS

Write a sentence that describes the location of the square counter.

The Object of the Game

Place your *GeoReflector* and a game piece or checker on a piece of paper so they match the drawing. Put your finger on the **image**. Is your hand in front of or behind the *GeoReflector?*

Place an **object** in front of the *GeoReflector.* Try to move the object without moving the image.

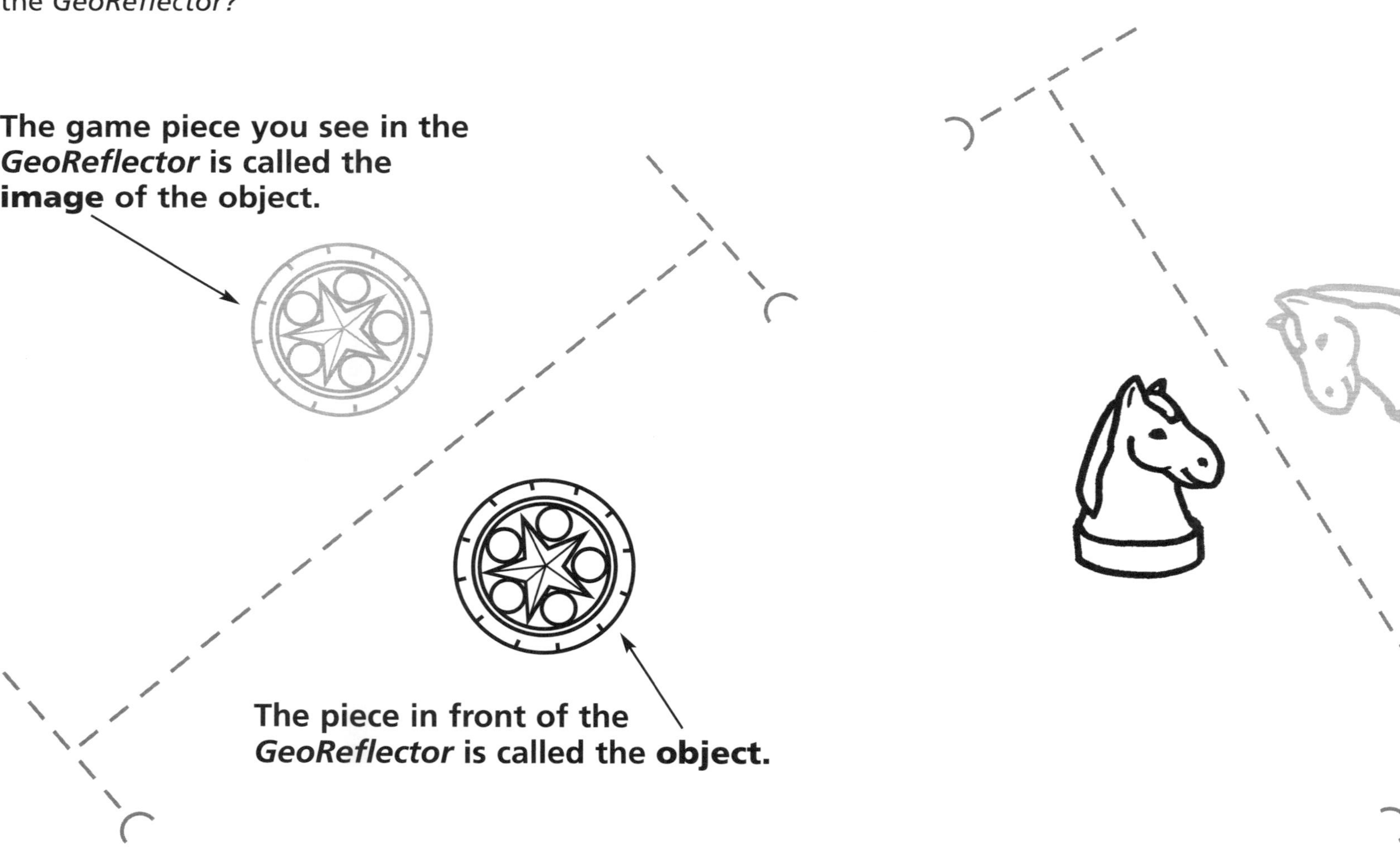

BONUS

What happens to the image when you move the object?

Penny, Penny

Stand the *GeoReflector* on the page as shown. Place one penny behind the *GeoReflector* and one in front of the *GeoReflector.* Move the **object** penny so its **image** covers the penny behind the *GeoReflector.* Do not move the *GeoReflector!* Draw a circle around the object penny.

Place two pennies in the spaces below. Place the *GeoReflector* on the page so the image of one penny covers—or **maps onto**—the other penny. Do not move the pennies! Draw a dashed line to show where you placed the *GeoReflector*.

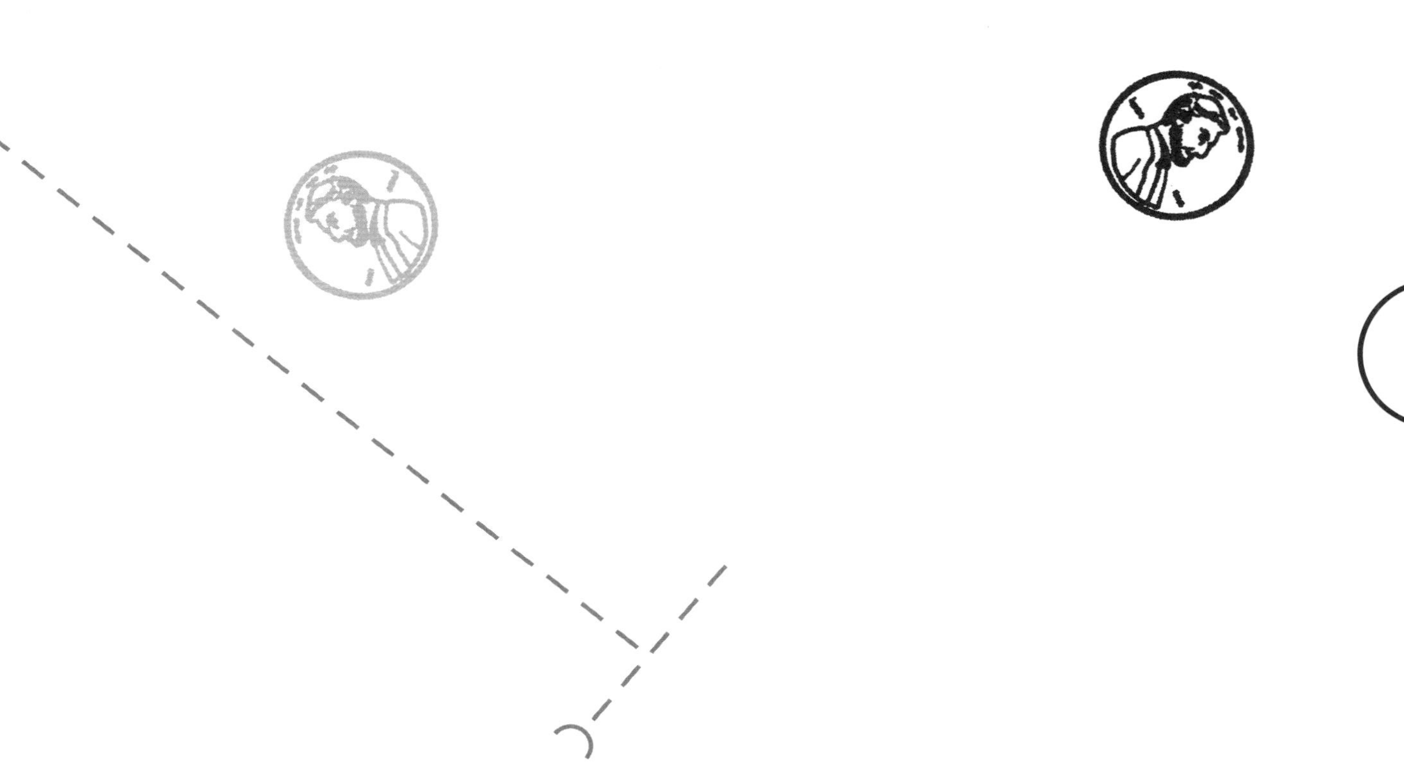

BONUS

What do you notice about the size and shape of an object and its image?

Be a Sport

Move your *GeoReflector* to make the gymnasts take turns vaulting the pommel horse.

Move your *GeoReflector* to make the hockey player chase the puck. Then, have the puck slide to the player!

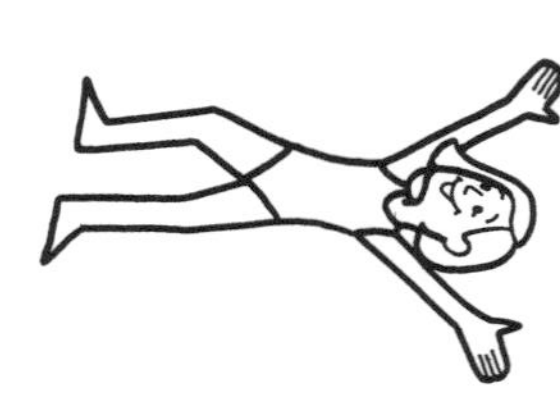

BONUS

Circle the gymnasts who did handstands on the pommel horse.

Making More or Less

Use these 6 hoops and the *GeoReflector* to make 12 hoops. Make 10 hoops, then 8 hoops.

Use these 6 squares and the *GeoReflector* to make 12 squares. Try to make 11, 10, 9, 8, and 7 squares.

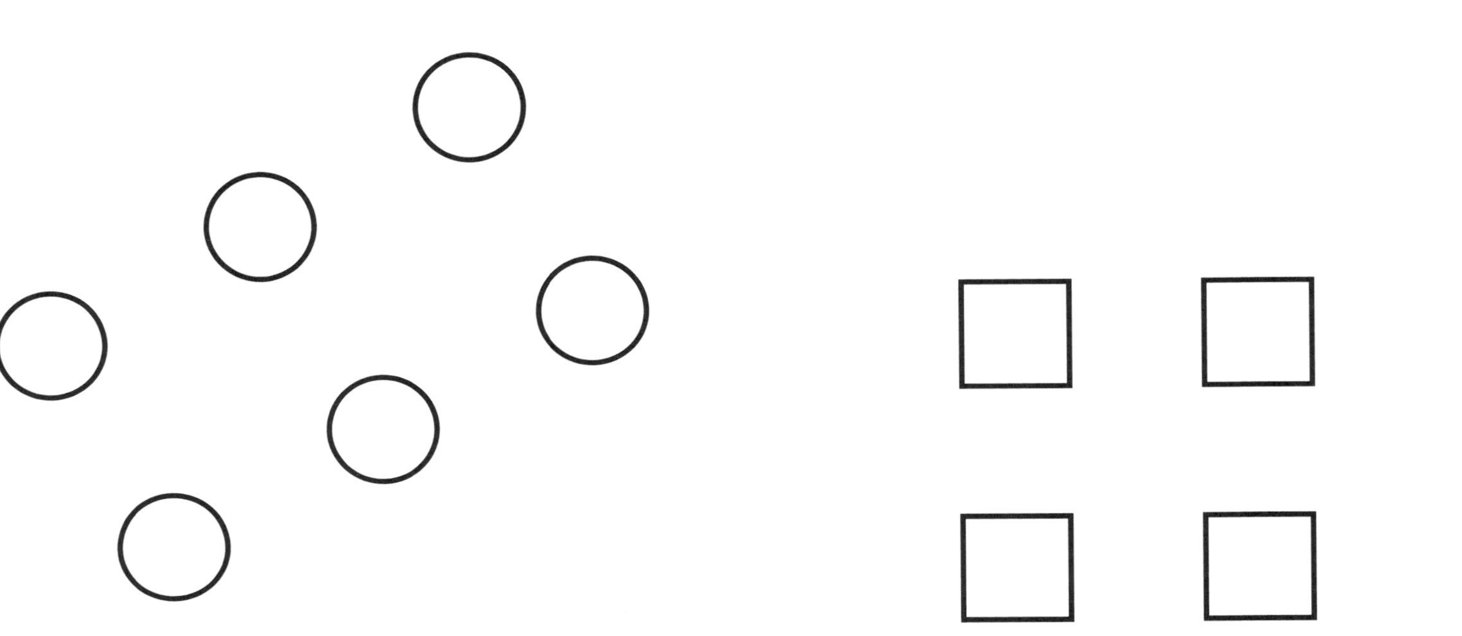

BONUS

Describe how you made 8 squares.

Magic Moments

Use these dominoes to make new dominoes.
Make a 2–3 domino ( ), a 1–5 domino ( ),
and a double-five domino ( ).

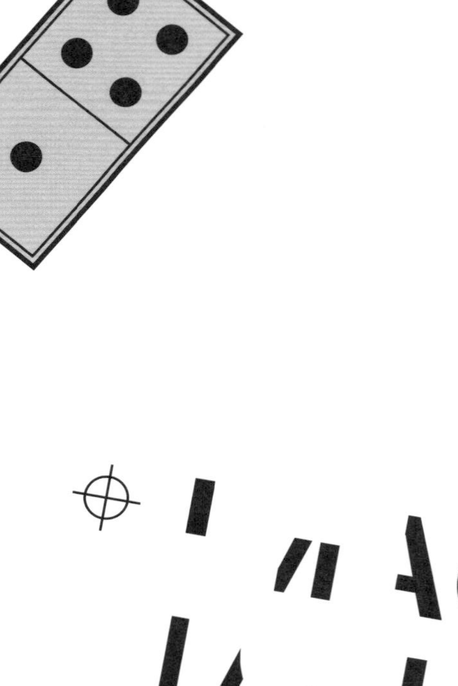

Use the *GeoReflector* to find the secret message.
HINT: Look at the "target points."

BONUS

Use the GeoReflector *to help you design your own secret message.*

Ready . . . Set . . . Draw

Place the *GeoReflector* on the **line of reflection**. Find the **reflection image** of the guitar in the *GeoReflector.* Trace the image while looking through the *GeoReflector* so there are two guitars on the page.

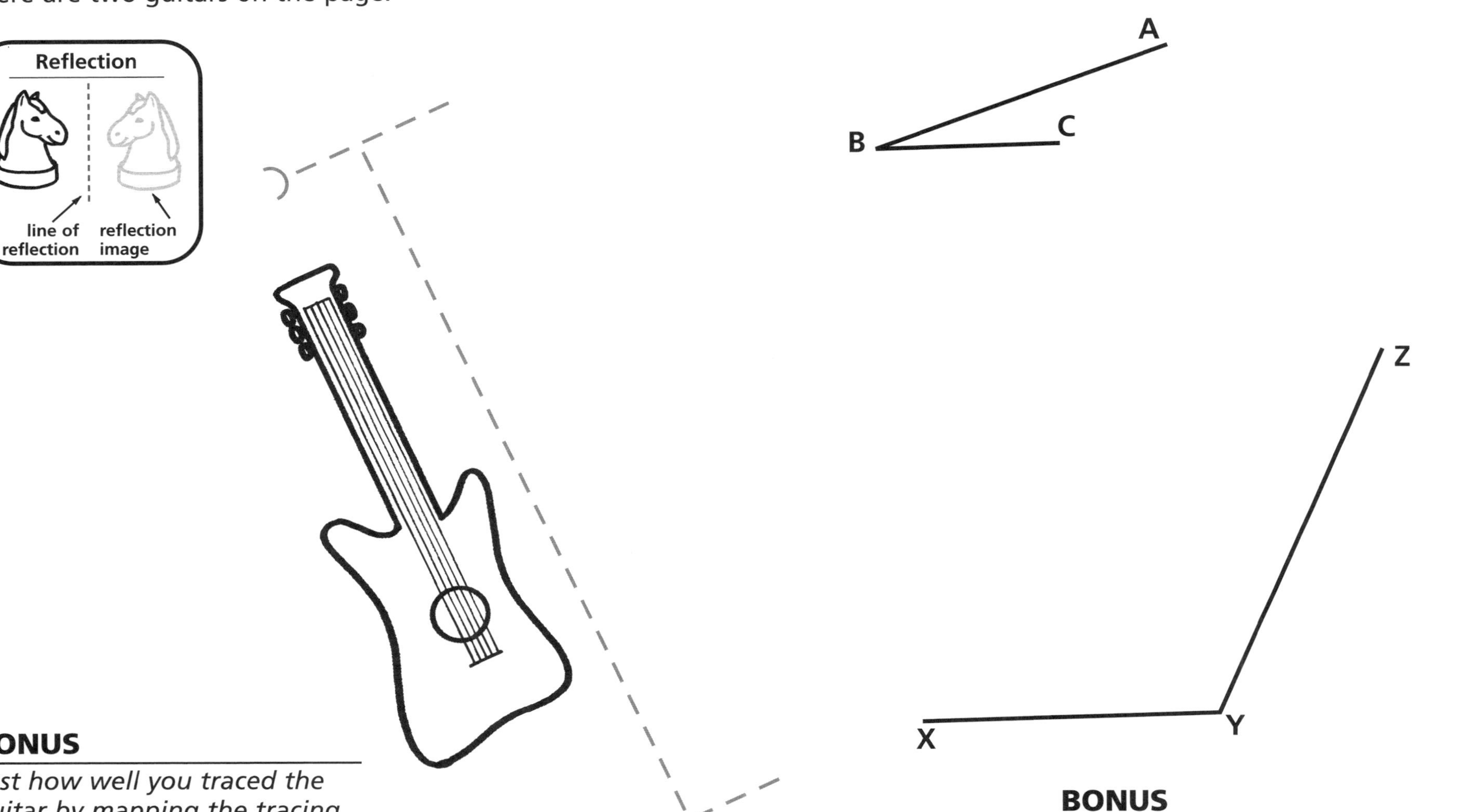

BONUS

Test how well you traced the guitar by mapping the tracing you made onto the original guitar. Use the same line of reflection.

Draw the image of lines $\overline{AB}$ and $\overline{BC}$ so that together the image and the object form a closed figure. Check your drawing by mapping the image onto the object.

BONUS

Form a closed figure using the GeoReflector *and lines* $\overline{XY}$ *and* $\overline{YZ}$*. Check your work.*

Taking Aim

Use the large arrow and the *GeoReflector* to make these combinations of arrows.

Place the *GeoReflector* so the reflected path of the golf ball eventually reaches the hole. Draw lines to show where the ball travels. Draw a dashed line where you placed the *GeoReflector.* The golf ball must stay within the walls.

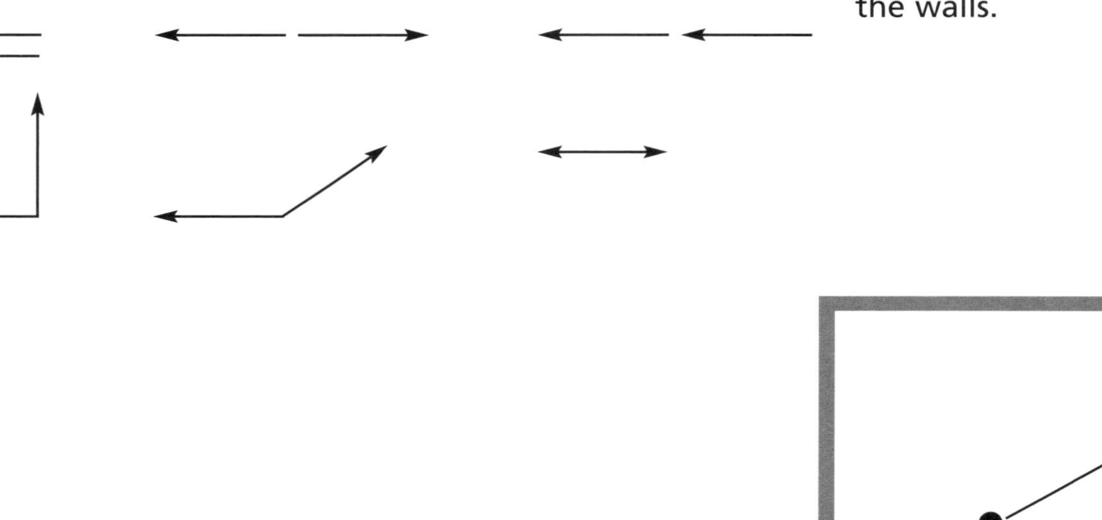

BONUS

Circle any combination that is not possible. Explain why each cannot be made.

The More the Merrier

Use this dancer to draw five other dancers.
Make a chorus line.

Draw a train that is two cars long, then four cars long.

BONUS

Check how well you drew the cars by mapping the last car you drew onto the first object car.

Design Codes

Use the *GeoReflector* to draw a star pattern that is two stars tall. Then make a design that is twice as long.

Make a design that turns a corner.

Making Faces

Place the *GeoReflector* on the dashed line labeled ℓ_1. Draw the image of the mask. Then place the *GeoReflector* on the dashed line labeled ℓ_2. Using your drawing as an object, draw the image you see. Move the *GeoReflector* and draw a fourth mask.

Place your *GeoReflector* on the dashed lines to draw five more faces like this one. Draw them in clockwise order.

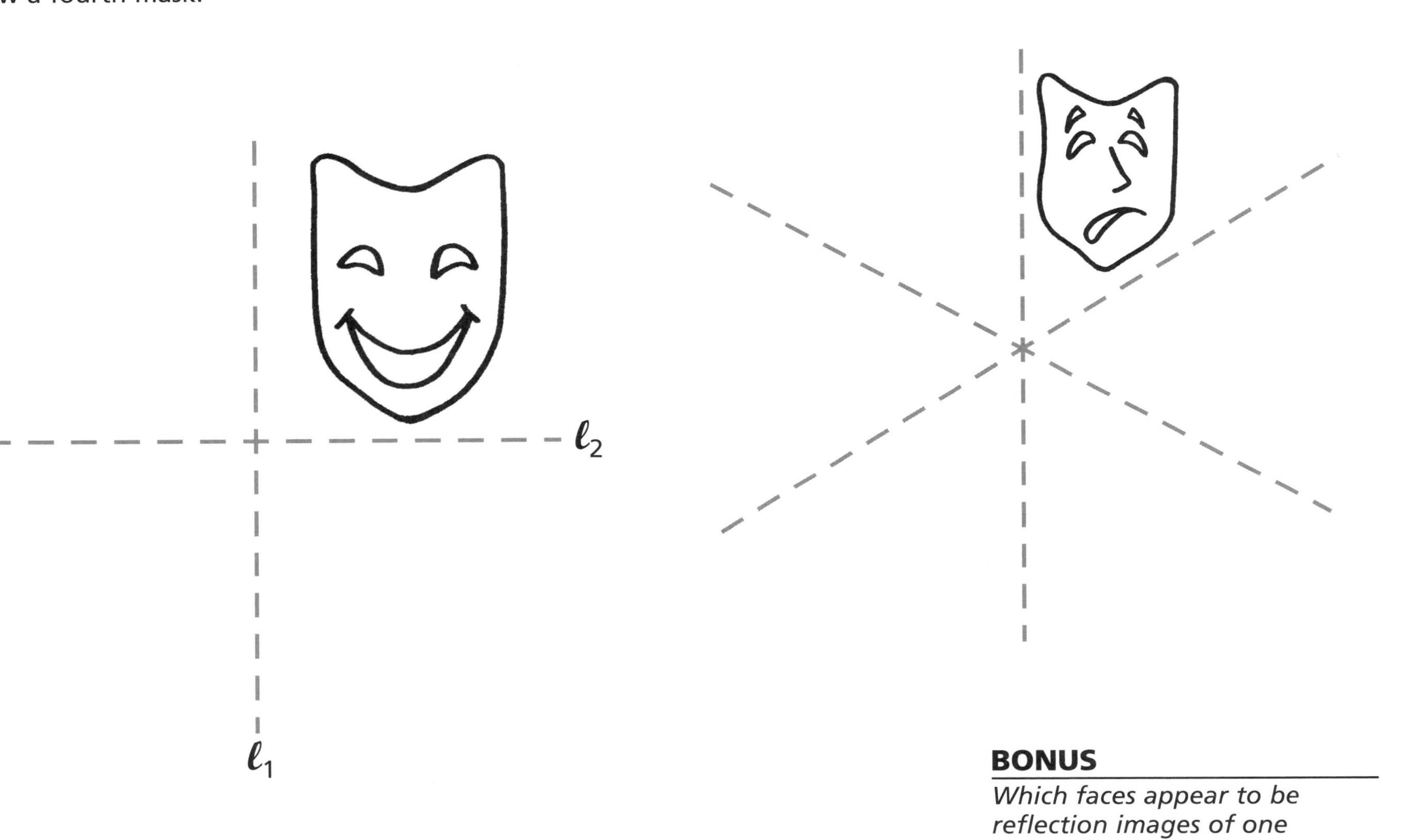

BONUS

Which faces appear to be reflection images of one another?

Pirate Eye

Place each of the parrots on the pirate's hand. Draw lines to show where you placed the *GeoReflector*.

This parrot is wearing a patch over its left eye. Stand the *GeoReflector* on the dashed lines to find three images of the parrot. Which image parrots are also wearing patches over their left eyes?

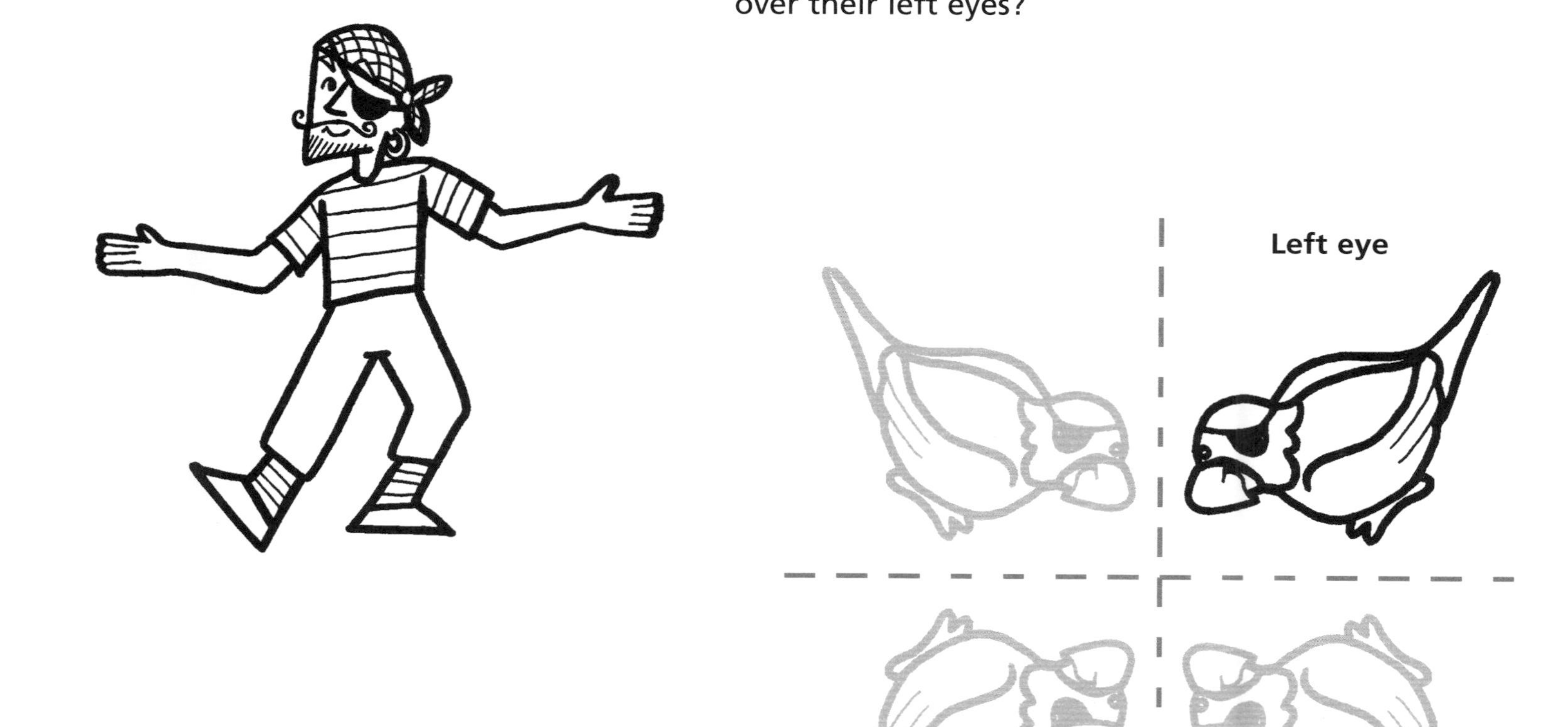

BONUS

Tell what you notice about the parrot's patched eye.

Motion & Image Manipulation

This section introduces students to the concept of manipulating the location of an image. Through slides and rotations, students discover how an image of an object may be positioned at some distance from the original image. Students learn what characteristics of the image are maintained, and what characteristics of the image are changed by the movement formally known as a translation. Students also explore the relationship of an object and its (flipped) reflection image. The activities in this section are self-assessing.

Pages 21 and 22 encourage students to compare the distance they move the *GeoReflector* with the distance the image moves. On page 21, students use a ruler to confirm their suspicions that the image moves twice as far as the *GeoReflector.* Students put that knowledge to use during the game on page 22 when they predict where to place the *GeoReflector* in order to locate the image at a specific point.

Page 23 revisits the motions used on page 21, and gives them geometric definitions of slide and rotation. It is important here that students can discriminate between object and image, and the act of moving an image from one location to another. It is not the object that slides or rotates, but rather its image. Students should recognize that the image is the same size and shape as the object, but it rests in a different location.

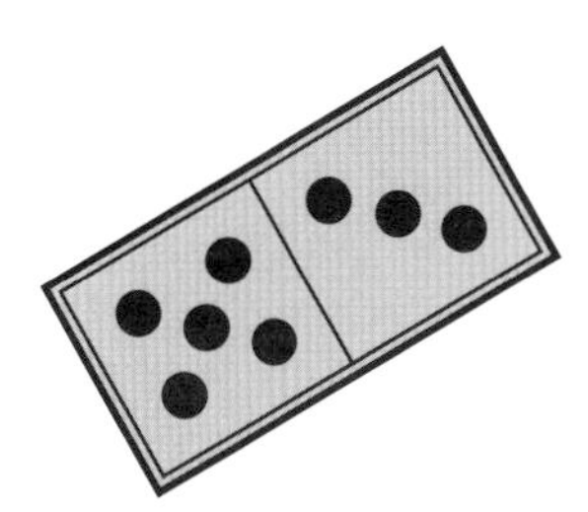

Page 24 strengthens students' ability to manipulate the *GeoReflector,* helping them control the location of an image.

Page 25 introduces students to another type of image manipulation: the flip. The asymmetric objects make clear the left-right or top-bottom reversal which occurs when an object is reflected in the *GeoReflector.* Point out to students that they can use cues such as the position of the musical instruments to tell if an image has been flipped or slid from one location to another.

Page 26 encourages students to look at the most ordinary objects to see if they can be flipped in one direction or another, and create an image that is exactly like the object. This activity presents an intuitive approach to symmetry. Symmetry will be explored further in later activities.

Page 27 invites students to be creative using flips, turns, and slides to create patterns. You might want to tell students that repeated patterns, called friezes, appear on many buildings as decorative elements. One way to make a repeated pattern is to manipulate images. Remind students to use what they know about flips and the left-right or top-bottom reversal as they try to identify patterns made by flips and slides.

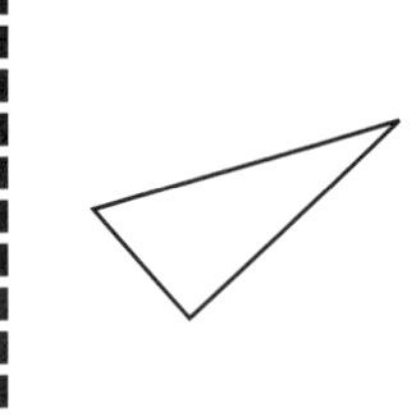

Page 28 offers a game in which students compete with one another to correctly identify flips, slides, and turns.

Slip Sliding Away

Place the *GeoReflector* just in front of each horseshoe. Next, slide the *GeoReflector* until the image of each horseshoe meets the goal post. Which moves farther, the image or the *GeoReflector?*

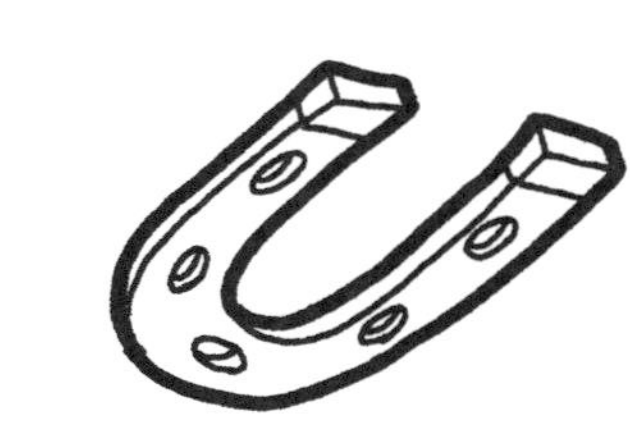

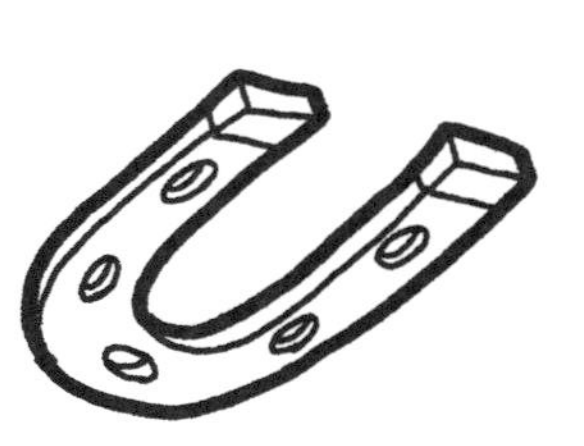

Place your *GeoReflector* on the dashed line near the domino. Draw the image of the domino. Next, slide an image of the domino to the top of the page by moving the *GeoReflector.* Draw the image of the domino and a dashed line along the edge of the *GeoReflector.*

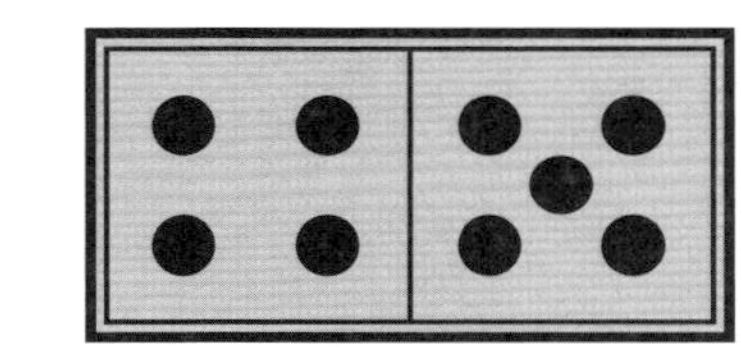

BONUS

Use a ruler to measure the distance the domino image moved and the distance the GeoReflector *moved. Which moved further? By how much?*

Image Toss

Try this game with a partner.

1. Estimate where to place the *GeoReflector* to toss a cream pie at the character's face.
2. Use the shaded pie as a practice turn.
3. Take turns placing the *GeoReflector* so the pies reach their target.
4. Once the *GeoReflector* is placed on the paper, it cannot be moved.
5. The player whose pie images come closest to the target wins.

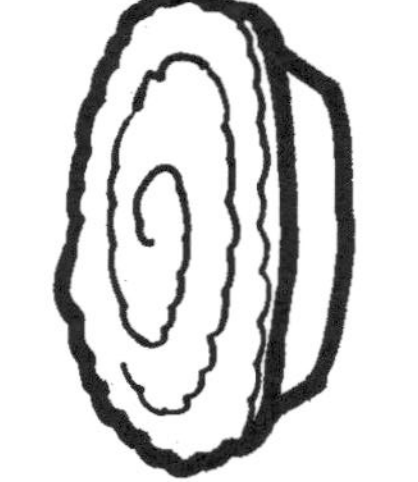
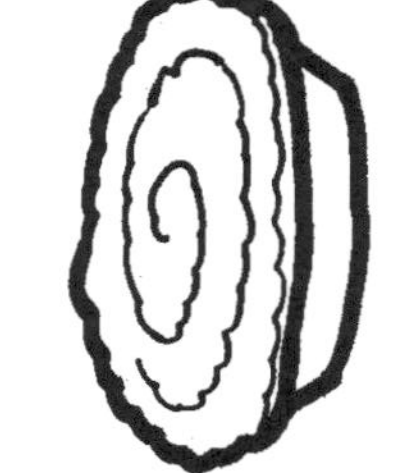

BONUS

Add your own pies to this game. Trade games with another partner pair. Play one another's Image Toss.

It's Your Turn

Move the *GeoReflector* away from the object domino to **slide** the domino image into the box at the middle of the page below. Draw the image of the domino in the box. Continue to move the *GeoReflector* to slide the object domino to the top box.

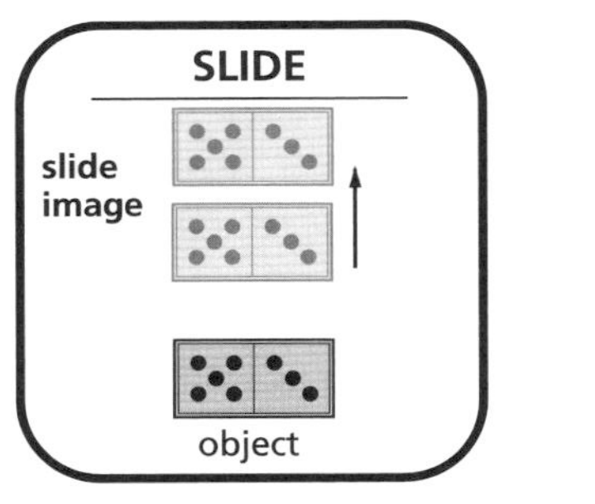

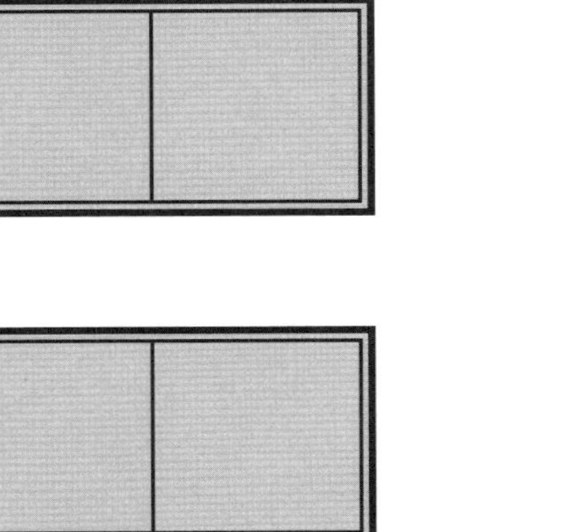

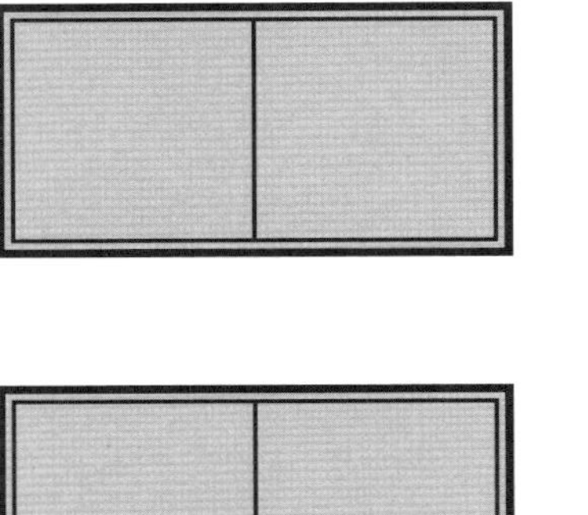

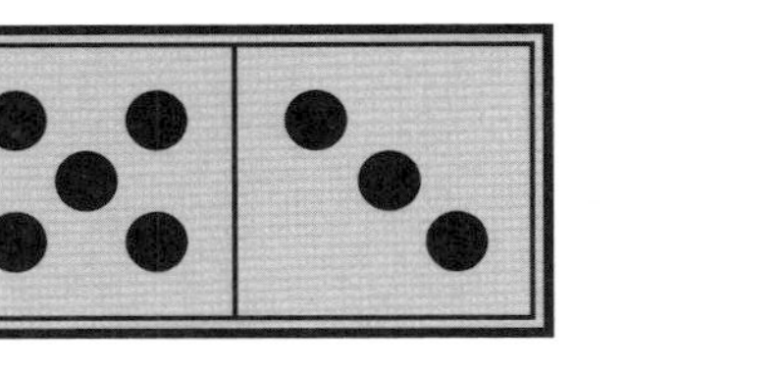

BONUS

How is the second image like the first image? How are they different from the object?

Use the *GeoReflector* to rotate the domino into the box. Draw the image of the domino in the box. Then, describe how you moved the *GeoReflector.*

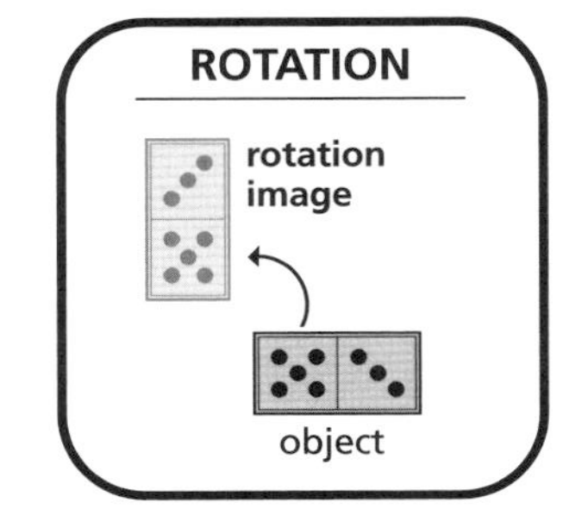

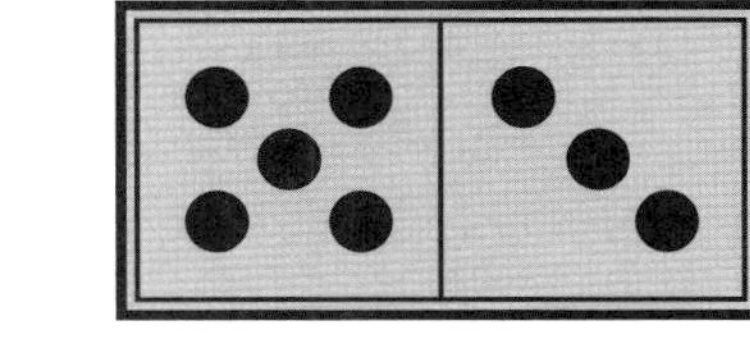

BONUS

How is the rotation image like the object? How is it different?

Rock Around the Clock

Use the *GeoReflector* to complete the clock face. Make the clock show 3 o'clock. Then make the clock show 6 o'clock.

BONUS

Move the image around the clock to show 7, 8, 9, 10, and 11 o'clock.

Move the image of the hand on the clock so that the time changes from noon to 4 o'clock. Draw a dashed line to show where the *GeoReflector* is to show 4 o'clock. Then change the time to 8 o'clock. Draw a dashed line to show where the *GeoReflector* is at 8 o'clock.

BONUS

Predict where the GeoReflector *has to be for the time to show 11 o'clock.*

Flip the Band

Use the *GeoReflector* to make a flipped image of the girl with the tambourine. Draw the image you make. Make at least two flips of the girl.

Complete the band. Make a singing trio (3) and a guitar quartet (4) using flips.

FLIP

image | object

BONUS

Describe how the flipped tambourine player is like the object player. How is she different?

Letter Flips

Predict where you should place the *GeoReflector* to get a dark object that flips to map over its gray image below.

Test your predictions. Draw dashed lines to show where you placed the *GeoReflector.*

BONUS

Which letters can be flipped sideways? Which letters can be flipped up and down?

Building Borders

Use your *GeoReflector* to make a decorative strip pattern at least 4 inches long. Use flips, slides, or a combination of both.

Use your *GeoReflector* to make a strip pattern that stretches diagonally across the page.

BONUS

Make a vertical pattern. Tell whether you used flips or slides. Can someone tell by looking at the pattern? Explain.

Slide, Flip, or Turn

Play this game with a partner.

1. Player 1 draws a straight line on the paper and a shape next to the line, like the shape shown.
2. Player 2 draws either a flip, slide, or turn of the shape. Decide which kind of image to draw by tossing two pennies: two heads—flip; two tails—slide; one head, one tail—turn.
3. Reverse roles and repeat steps 1 and 2.

Either player can challenge the other's drawing, checking the drawing with the *GeoReflector.* If the drawing is correct, the first player earns a point. If the drawing is incorrect, the challenger earns a point. The first player to 10 points wins.

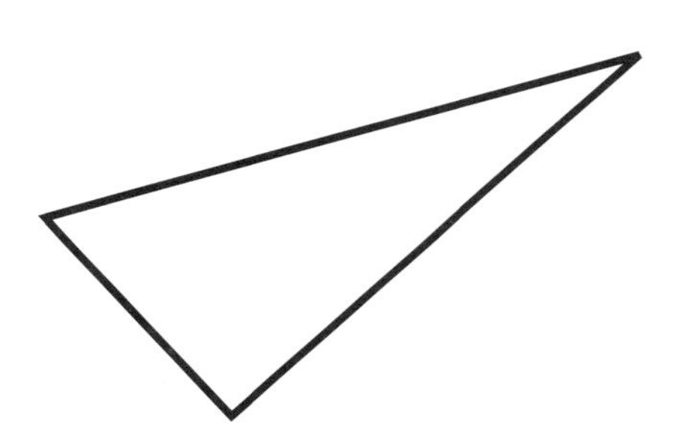

Symmetry and Congruence

This section explores symmetry and congruence. The activities are more structured than previous activities. A few geometric terms—symmetry, line of symmetry, corresponding points, and congruent images—are defined for students. As with previous sections, these activities build upon one another.

Pages 31 and 32 put forth the notion that many common shapes and items are symmetric. Remind students that as they look through the *GeoReflector,* the object lines will appear black while image lines appear blue. Help students understand that once the image maps exactly onto the object, they have found a line of symmetry, and the *GeoReflector* is resting upon that line. Encourage students to search out other familiar symbols, such as trademarks and logos, and have them test several of these for symmetry.

Page 33 is a direct extension of an earlier activity. Have students recall which letters they could flip without changing the look of the letter. Suggest they compare that list of letters to the symmetric letters presented in this activity.

Page 34 introduces congruent objects by asking students to find a line of symmetry which exists between two separate objects rather than within one object. The exercise is self-evaluating. Once students see that the object and image map onto one another, they have found the line of symmetry.

Page 35 introduces the concept of symmetry with common, basic geometric shapes. Some of the regular many-sided figures may intrigue students because they have several lines of symmetry. Encourage students always to look for more than one or two lines of symmetry when they are testing a shape for symmetry.

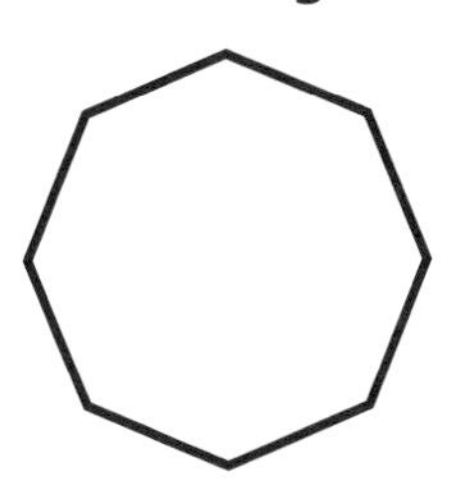

Page 36 and 37 formalize the relationship between image and object by directing students to identify points on an image which correspond to specific points on an object. Here they learn that object points closer to the line of reflection correspond to image points closer to the line of reflection. Their explorations lead them to conclude that corresponding points are always at equal distances from the line of symmetry.

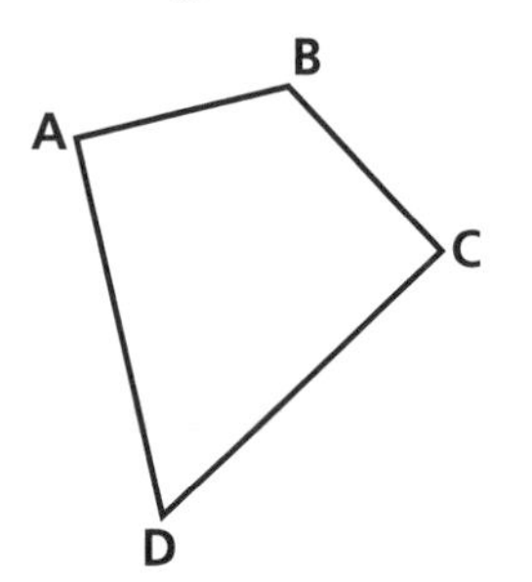

Pages 38, 39, and 40 lead students to discover a way to test whether shapes are congruent. Congruent shapes are identical. The activity on page 40 requires tracing paper as an alternative for testing whether objects are, indeed, congruent.

Pages 41 and 42 are designed to show students how symmetry can be used to create patterns and illusions. On page 41, students are presented with an all-over design and asked to find congruent images or the line of symmetry. On page 42, students complete a symmetric scene that relies on the placement of the *GeoReflector*.

One of these Halves

Place the *GeoReflector* so the image of half the figure fits exactly onto the other half of the figure.

Draw a dashed line to show where you placed the *GeoReflector.* This is the **line of symmetry**.

BONUS

Do you think you can reflect the image of a rectangle onto itself? Draw a rectangle and test your pattern.

Line Test

Using your *GeoReflector* as a guide, draw the line of symmetry for each object below.

BONUS

Which object above does not have a line of symmetry?

Letter Symmetry

Find and draw the lines of symmetry for each of the letters below. Some of these letters have more than one line of symmetry.

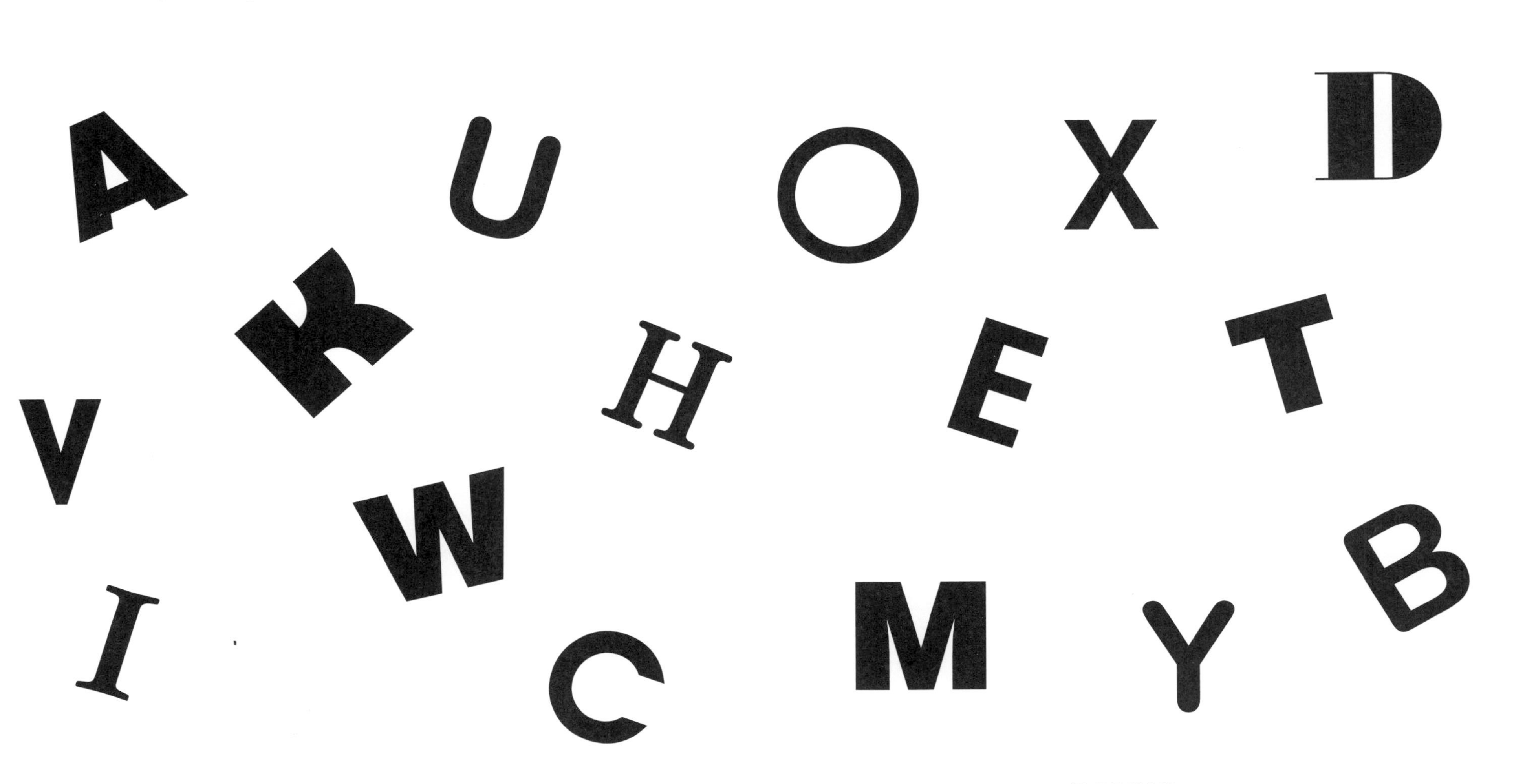

BONUS

Tell how you could use letter symmetry to develop a secret message.

Pair the Pairs

Use your *GeoReflector* to find the line of symmetry for each pair of same-size circles. Draw the line of symmetry. Color each symmetric pair the same color.

Find the line of symmetry for the pairs of squares. Draw the line of symmetry. Color each symmetric pair the same color.

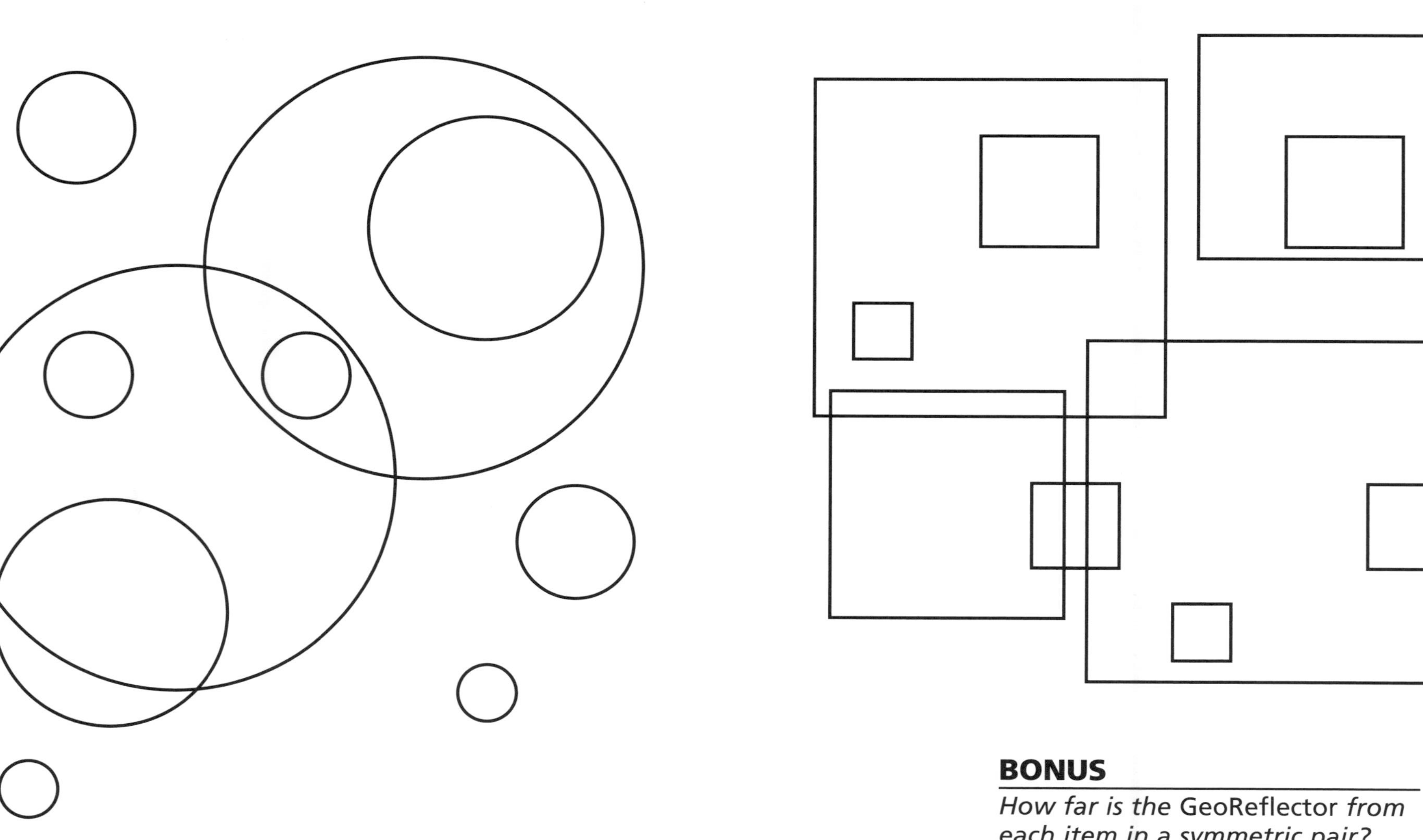

BONUS

How far is the GeoReflector *from each item in a symmetric pair? Use a ruler to test your prediction.*

Symmetry Side Show

Find and draw two lines of symmetry for three of these figures.

Find and draw at least three lines of symmetry for each of these figures.

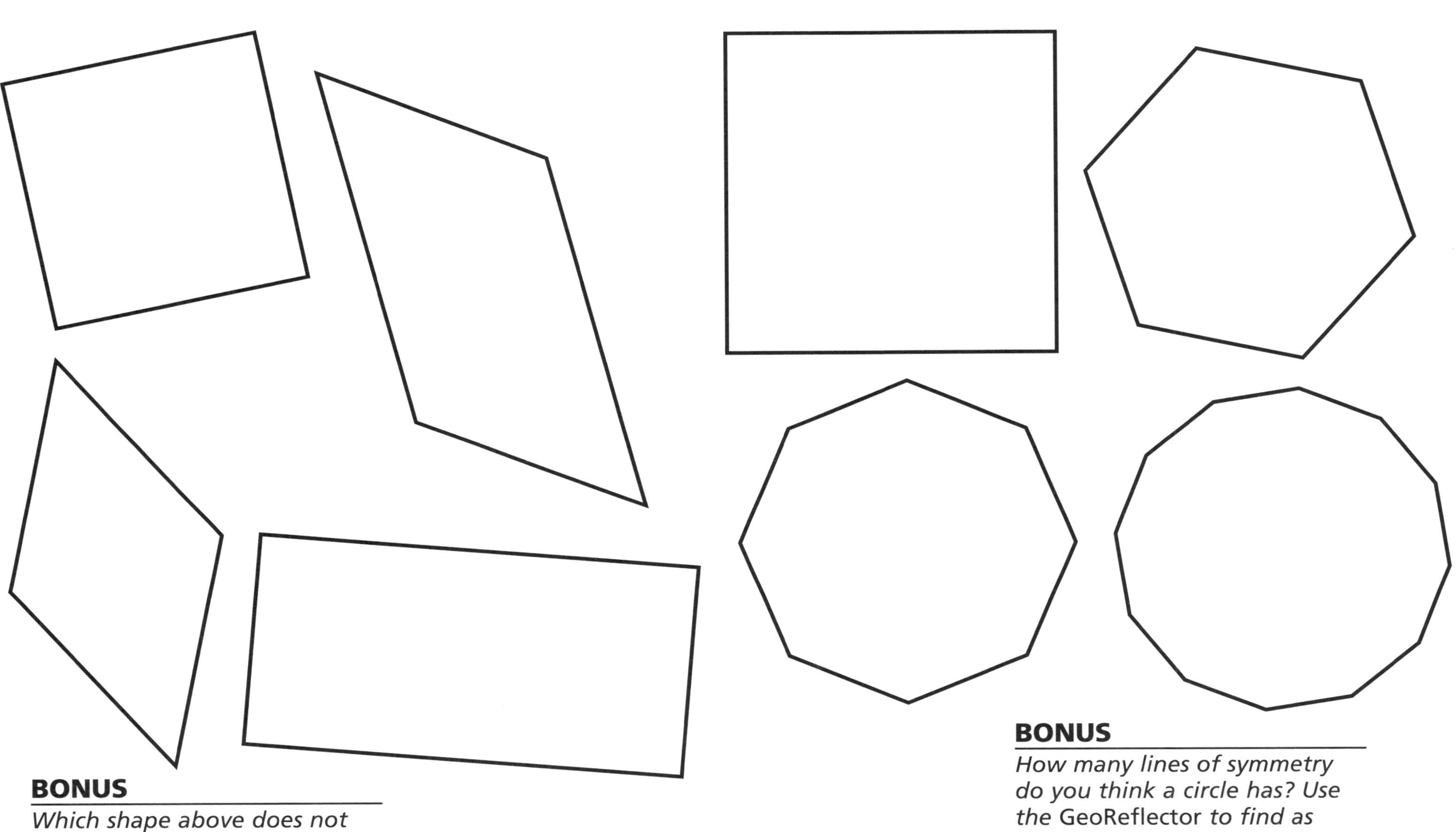

BONUS

Which shape above does not have lines of symmetry?

BONUS

How many lines of symmetry do you think a circle has? Use the GeoReflector *to find as many as you can.*

A Prime Example

Draw the line of symmetry for each pair of figures.
Label the corresponding points

Draw the line of symmetry for these pairs of figures.
List the corresponding points in the chart.

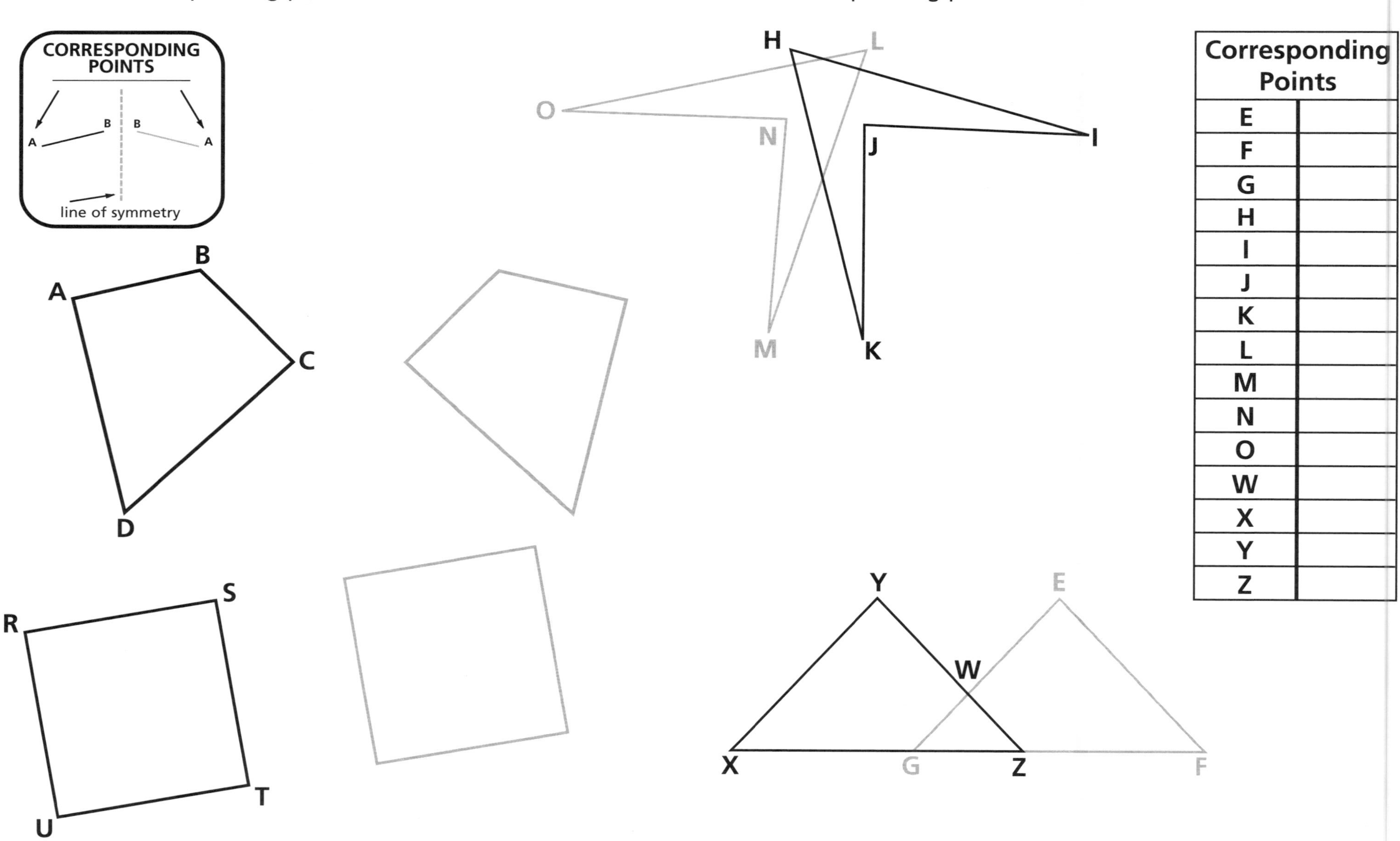

Corresponding Points	
E	
F	
G	
H	
I	
J	
K	
L	
M	
N	
O	
W	
X	
Y	
Z	

Correspondence School

For each pair of figures below:

1. Draw the line of symmetry.
2. Label the corresponding points.
3. Measure the shortest distance from each point to the line of symmetry.
4. Measure the distance between each pair of corresponding points.

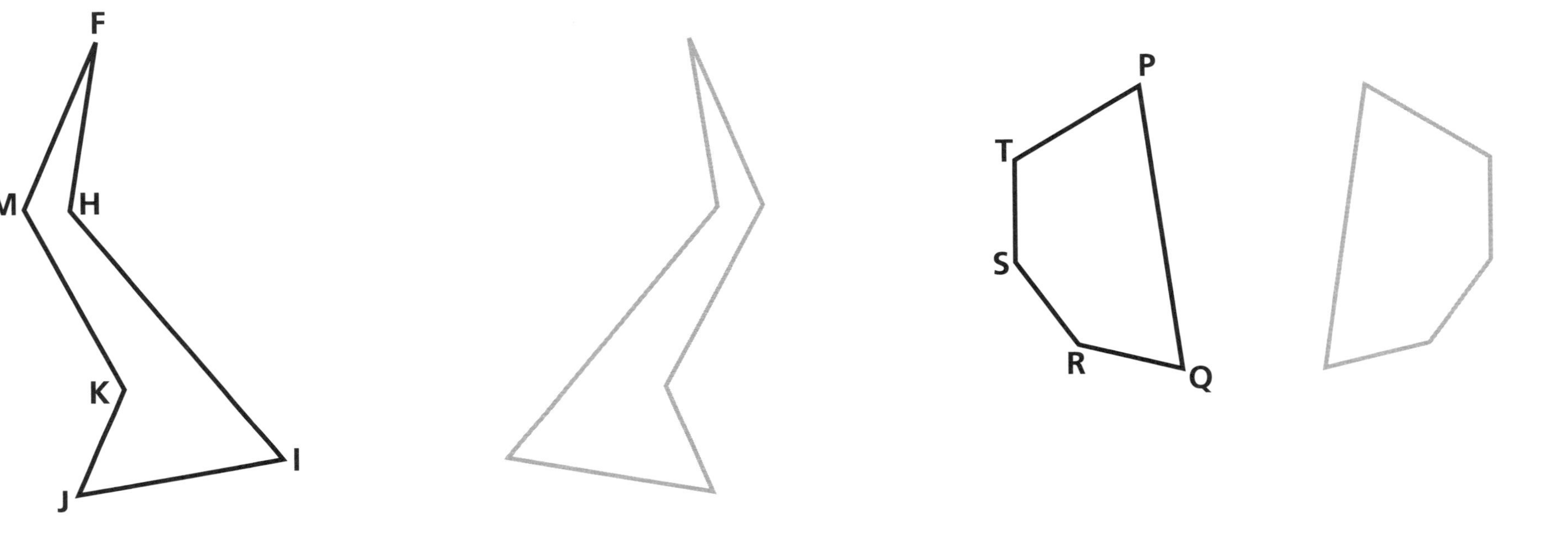

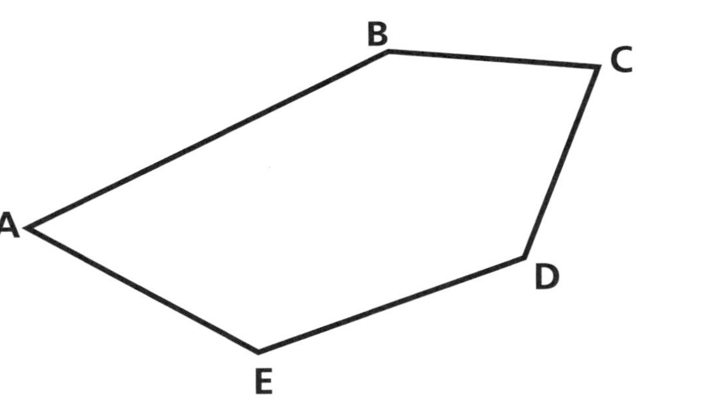

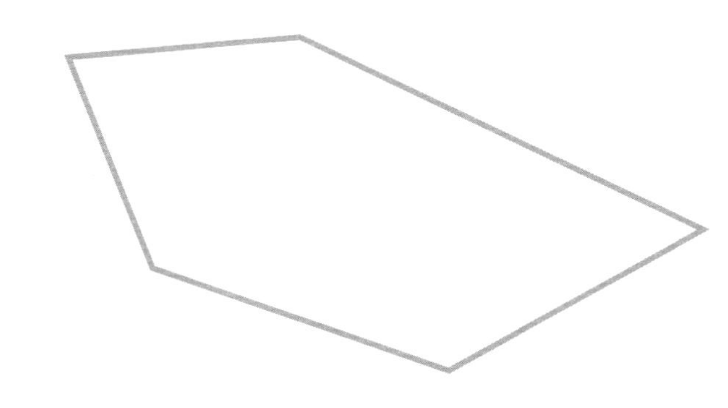

BONUS

Write a sentence describing what you notice about the distance between corresponding points, and the distance from the points to the line of symmetry.

Spot, the Matching Dog

Find the pair of exactly matching dogs.

CONGRUENT

line of reflection

A D

C B F E

If two figures are the same size and shape, they are **congruent.** Congruent is another word for **identical.**

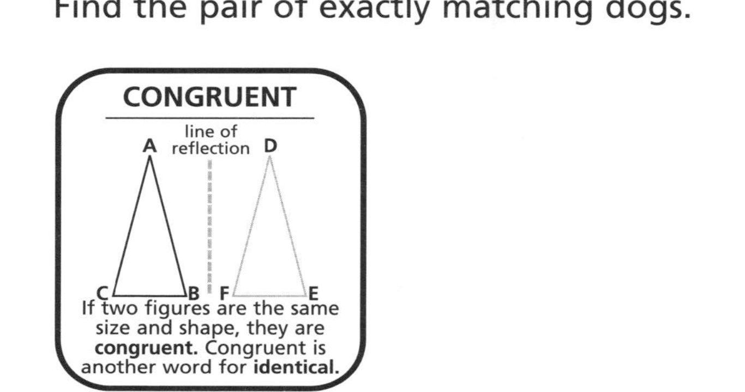

Find congruent pairs of items. Color congruent pairs the same color.

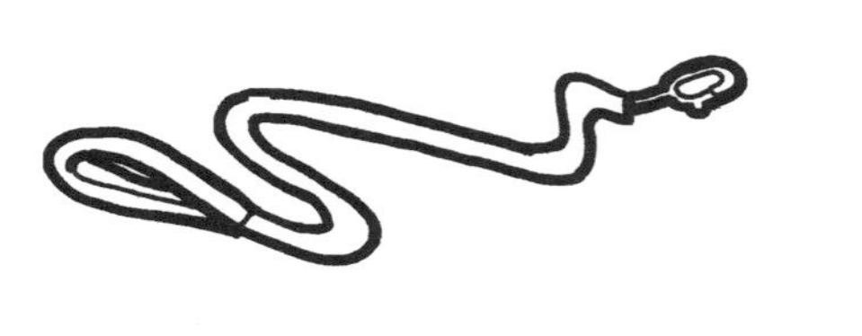

BONUS

Tell how you can use reflections and images to find matching pairs.

Match-Making

1. Predict which of these pairs of figures are not congruent.
2. Use the *GeoReflector* to test your prediction.

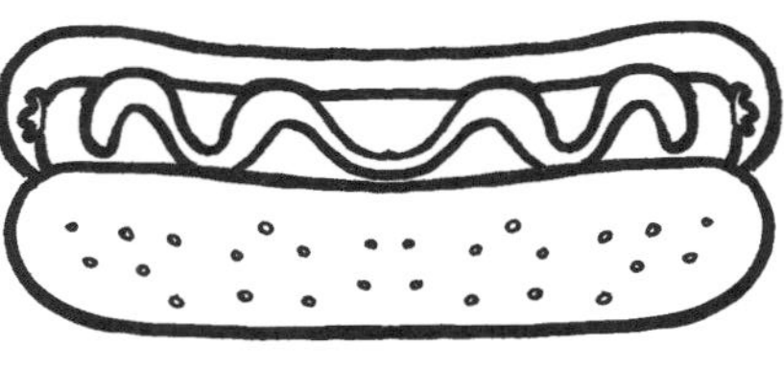

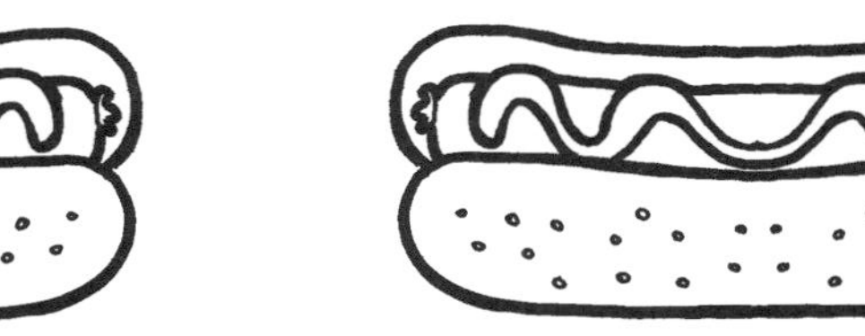

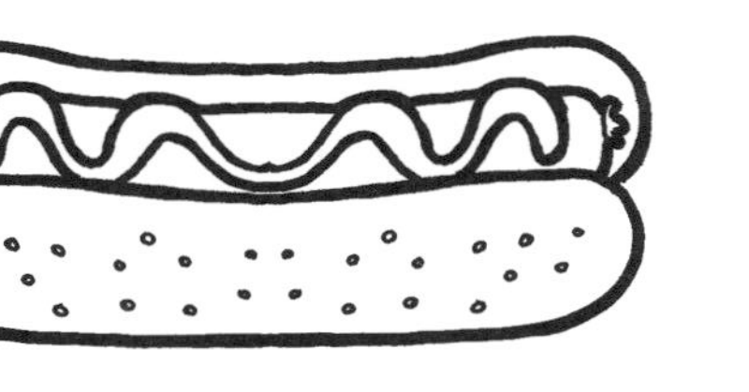

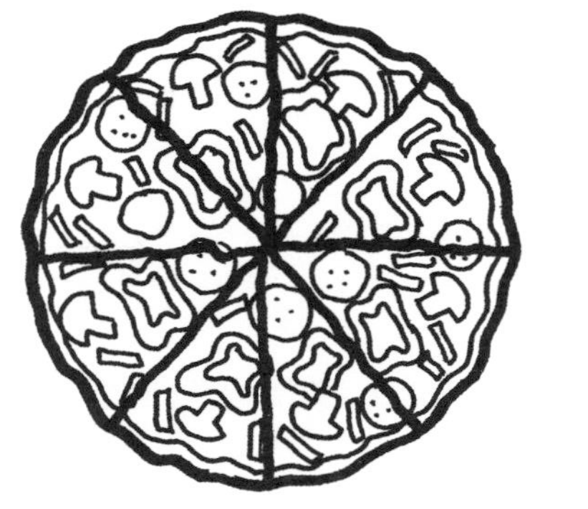

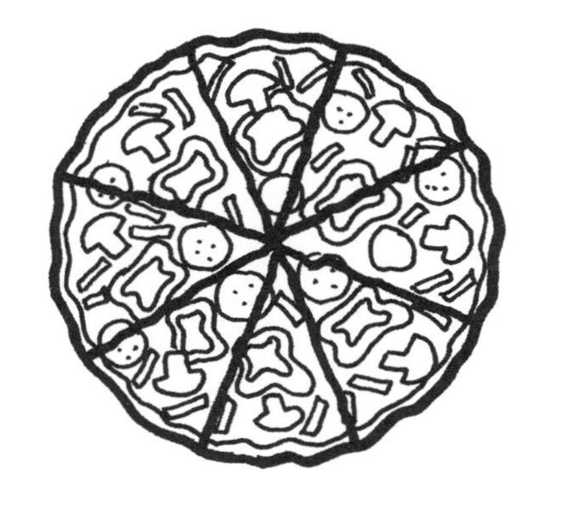

BONUS

Trace one of the snack foods onto tracing paper. Explain how you can use the tracing paper to find out if the snack food pairs are congruent.

The Terrific Triangles

Find out if the Flying Triangles are congruent. Use the *GeoReflector* and tracing paper. Color congruent triangles the same color.

Find out if the Tumbling Triangles are congruent. Use the *GeoReflector* and tracing paper. Color any congruent triangles you find the same color.

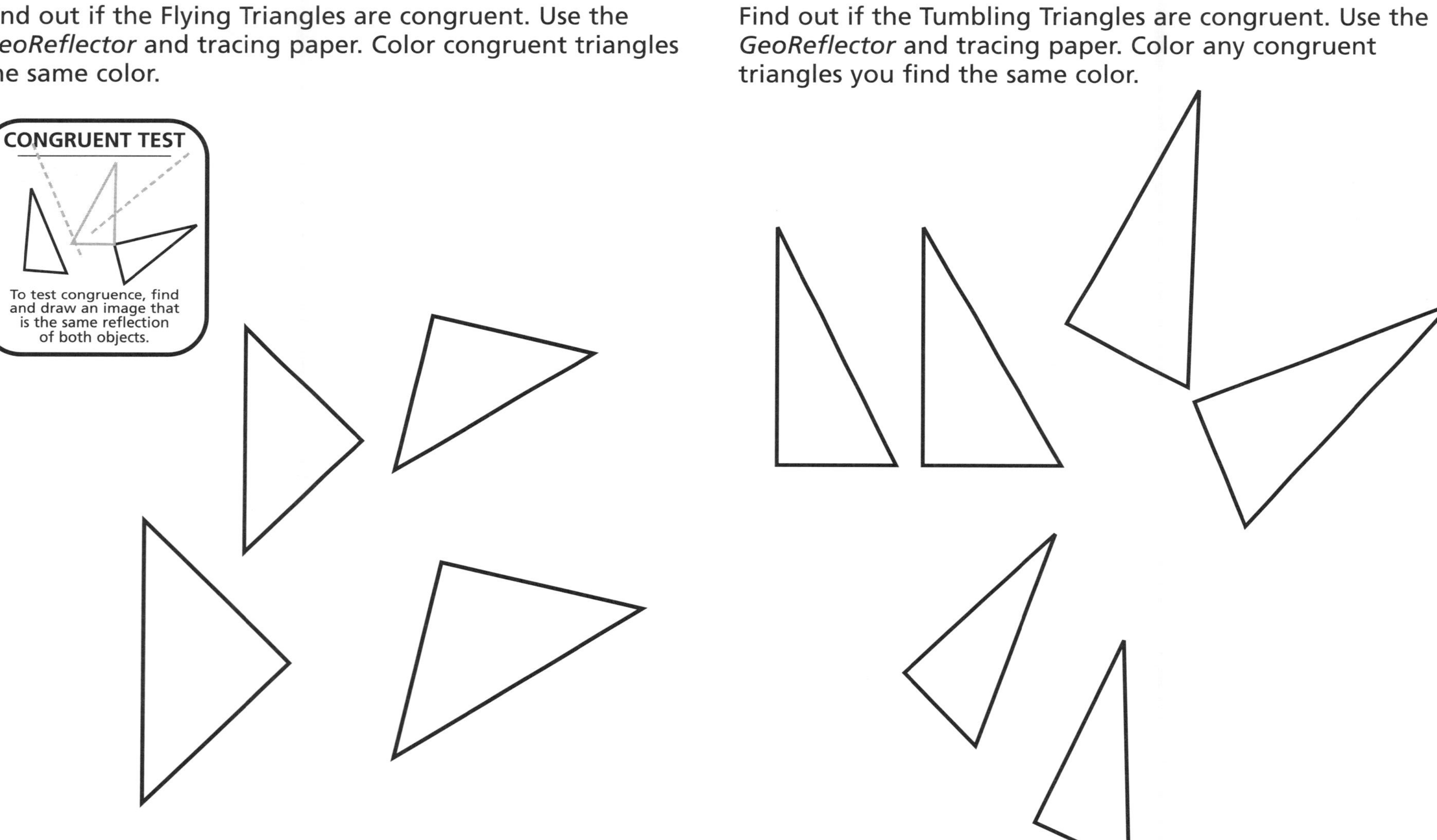

BONUS

Write directions that tell how you found out if the triangles were congruent.

Line Search

Find where to place the *GeoReflector* to show that Butterfly A and Butterfly B are each symmetric. Use tracing paper to find out if Butterfly C and Butterfly D are symmetric.

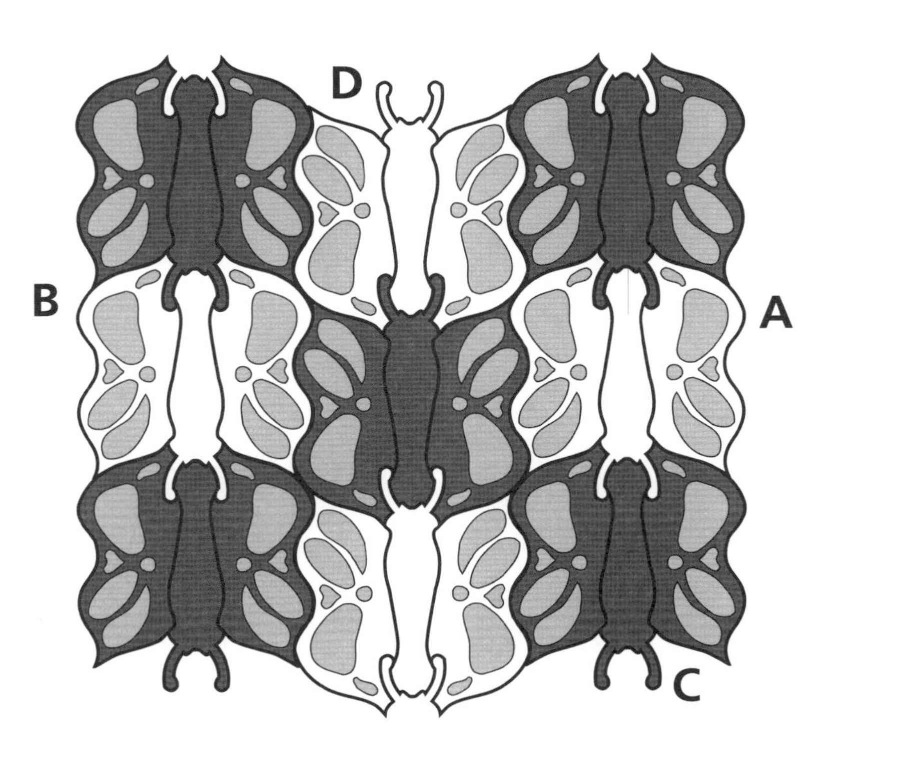

Find two dogs in the pack that match the reflected image of the lone dog. Color the three matching dogs.

Complete the Picture

Use your *GeoReflector* to complete the picture.

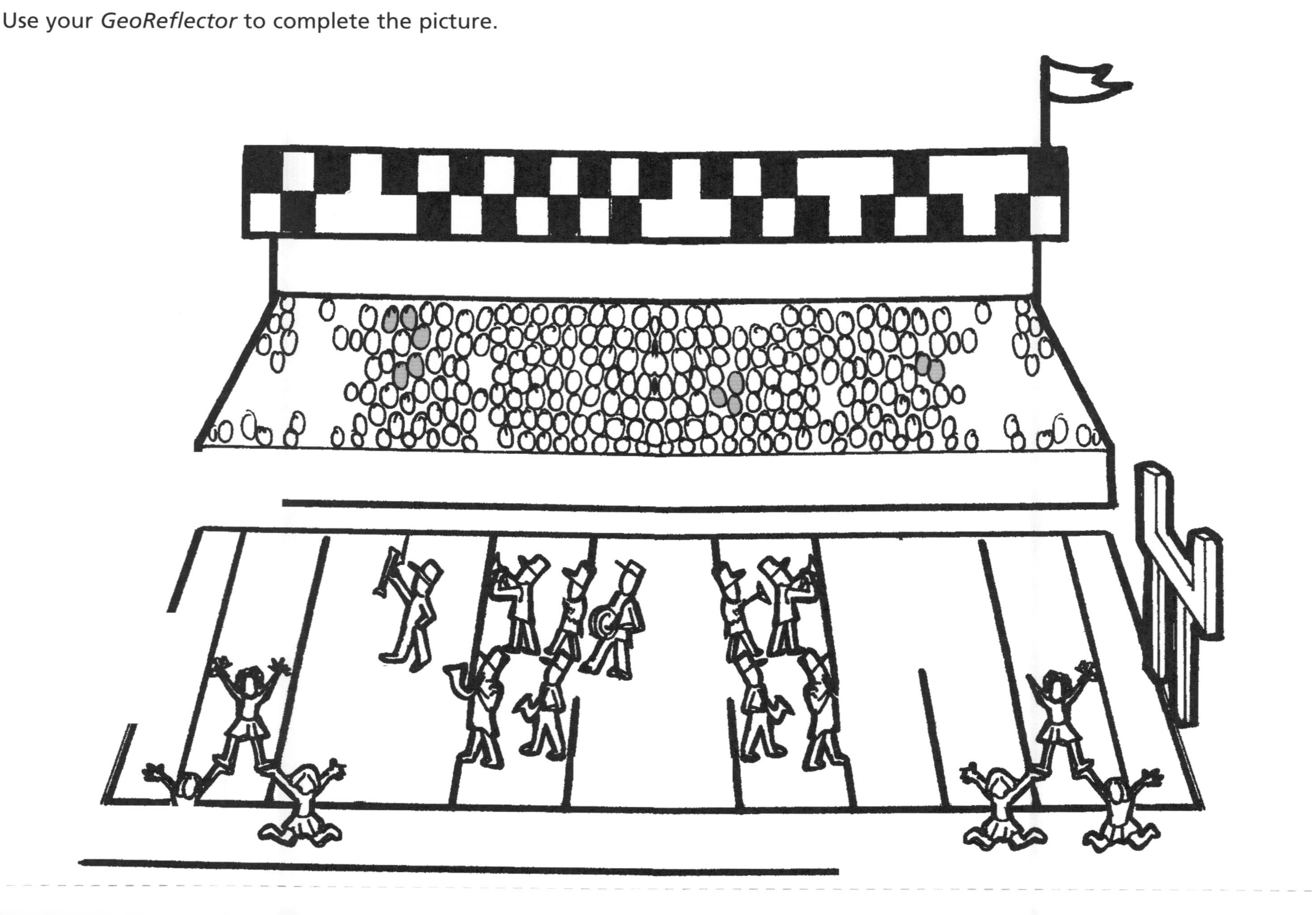

Geometry & the *GeoReflector*

This section introduces students to simple but important geometric constructions. The *GeoReflector* makes short work of several complex geometric constructions. It allows students to focus on the results of the construction rather than the method for completing the construction itself.

For best results, remind students to work carefully. Students should use a well-sharpened pencil and the *GeoReflector's* beveled edge as they draw lines of symmetry, angle bisectors, medians, and altitudes. Remind them that they have already discovered that a small movement in the *GeoReflector* results in an image movement of twice the distance.

This section assumes students have some knowledge of geometry-related terms (vertex, triangle, rectangle), but does not address the distinction between lines and line segments. Some activities in this section require the use of a compass or other circle-making device. Compass activities are marked with a small compass icon in the corner of each page.

Page 45 presents students with a method for testing whether pairs of lines are parallel. Students are encouraged to develop their own test criteria, for example, "If the lines are parallel, the image always maps exactly onto the line."

Pages 46 and 47 show students how to use the *GeoReflector* to draw perpendicular lines. Students also test to find that lines perpendicular to the same line are parallel to one another. Finally, students practice drawing a perpendic ular to a line through points both on and off the line.

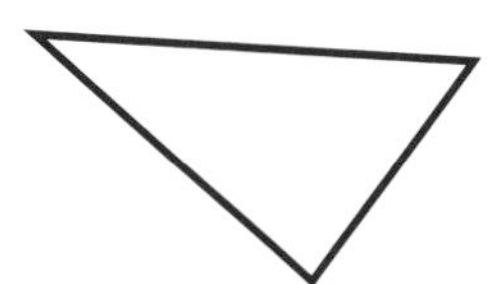

Page 48 Students construct a perpendicular to a line (side of a triangle) through a point not on the line (the opposite vertex). By exploration, students find that the altitudes of a triangle intersect at a single point. Encourage students to test their conclusions with several triangles of their own.

Page 49 relies on students' skill in mapping to have them find the midpoint of a line. This activity helps prepare students for constructing perpendicular bisectors, as well as to introduce the medians of a triangle. Students find that the medians of a triangle meet at a single point.

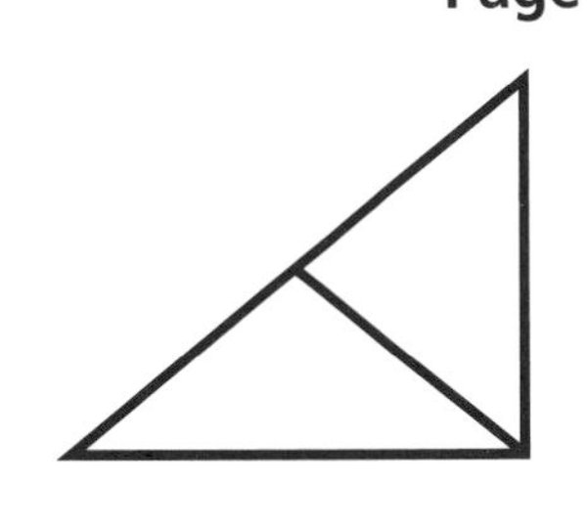

Pages 50 and 51 allow students to practice constructing perpendicular bisectors. Students should conclude that the perpendicular bisectors of a triangle meet at a single point. Encourage students to test their conclusions with triangles of their own.

Pages 52 and 53 are related exercises concerning the construction of the outcircle. An outcircle is a circle whose center is at the intersection of the perpendicular bisectors of the triangles' sides, and passes through each of the triangles' vertices.

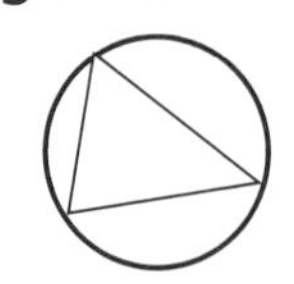

Pages 54, 55 and 56 introduce students to the construction of angle bisectors using the *GeoReflector*, and the relationship between angle bisectors and lines of symmetry for some geometric shapes. Angle bisectors of a triangle meet as a single point and use this point as the center of a circle which touches all sides of the triangle. Students also find that the perpendicular bisectors and angle bisectors of some regular polygons meet at a single point.

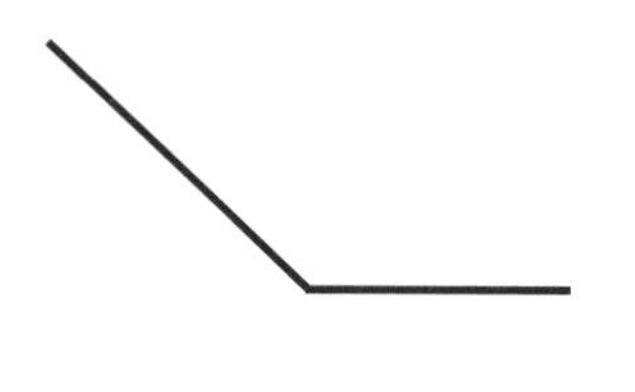

Pages 57, 58, 59, and 60 require students to combine elements of what they have learned to accomplish specific objectives. Students draw circles that pass through specific points, construct squares or octagons within circles, construct isosceles and equilateral triangles, and choose and construct a triangle whose incircle and outcircle have the same center. While each of these activities is self-assessing, sample solutions are provided at the back of the book.

Parallel Tests

Place the *GeoReflector* across one pair of lines. Turn the *GeoReflector* until the image of the lines maps onto the lines themselves. What do you notice about the images of the parallel lines?

Use what you know about the images of parallel lines to find parallel lines in these figures. Circle any sets of parallel lines you find.

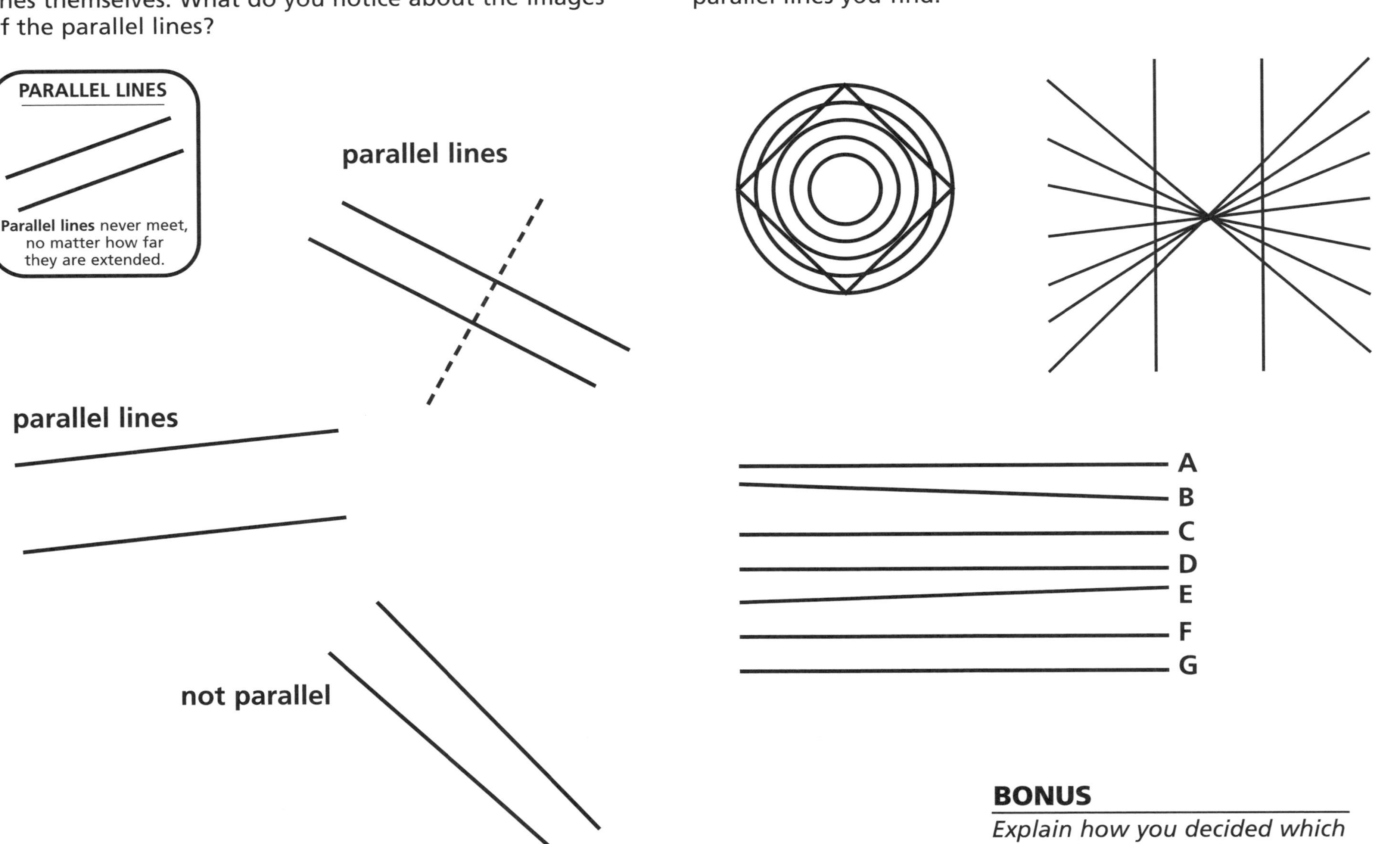

BONUS

Explain how you decided which figures had parallel lines.

Right and Perpendicular

Place the *GeoReflector* anywhere along line ℓ so that the image of ℓ maps onto itself. Draw a line along the edge of the *GeoReflector,* and label it m. Line m and line ℓ are **perpendicular.**

PERPENDICULAR LINES

90°

Perpendicular lines form a 90° or right angle.

Draw two lines, each perpendicular to line ℓ_1. Label the lines m and n. Place the *GeoReflector* so line m maps onto itself. What happens to line n? Repeat this activity with line ℓ_2. Do you get the same results?

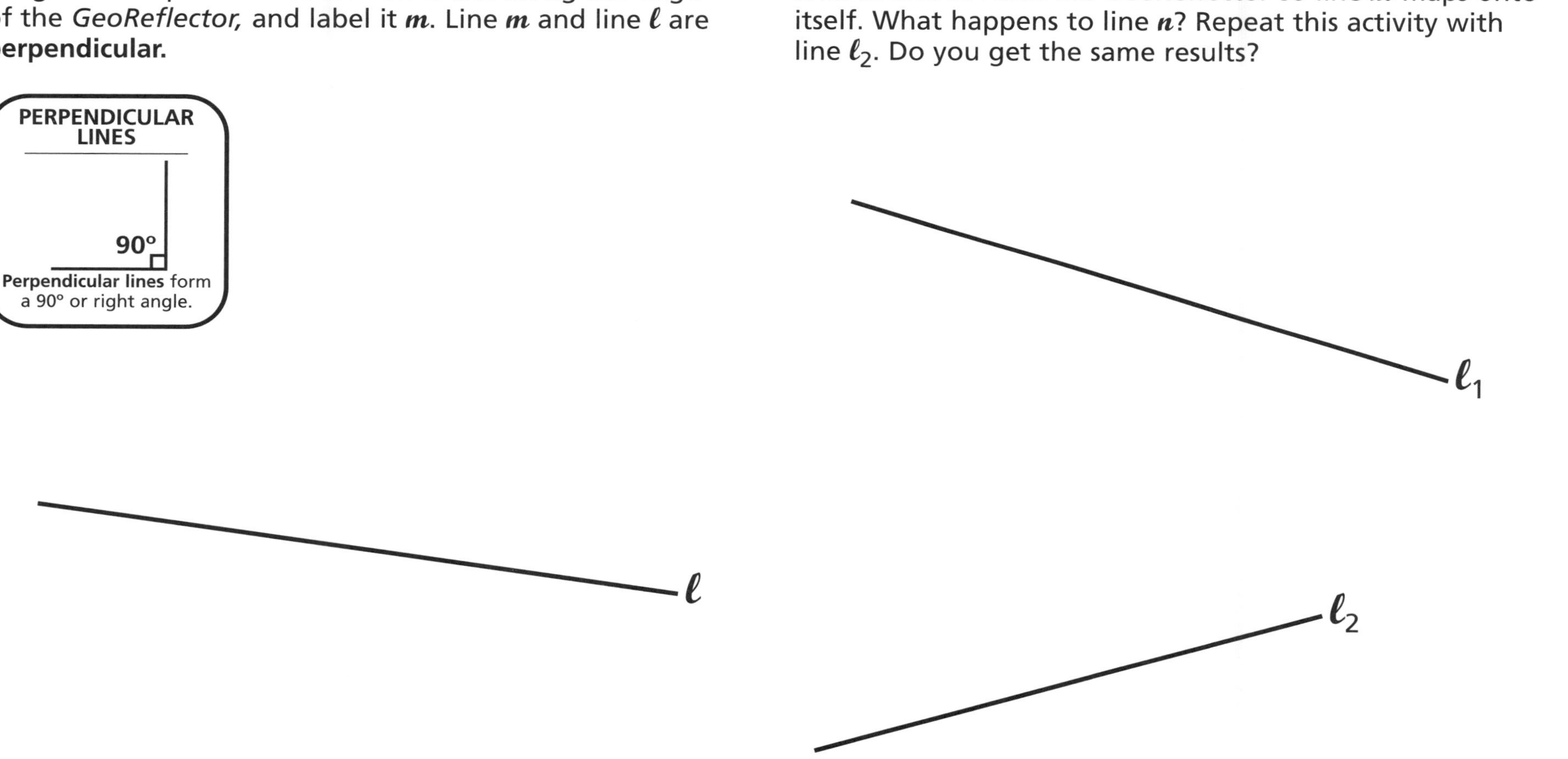

BONUS

How many lines can you draw that are perpendicular to line ℓ?

A and *m* Construction

Use your *GeoReflector.* Draw a line perpendicular to ℓ that passes through point A. Then draw a line perpendicular to ℓ that passes through point B.

Draw a rectangle that has $\overline{AB}$ as its base. Draw a square that has $\overline{FG}$ as its base.

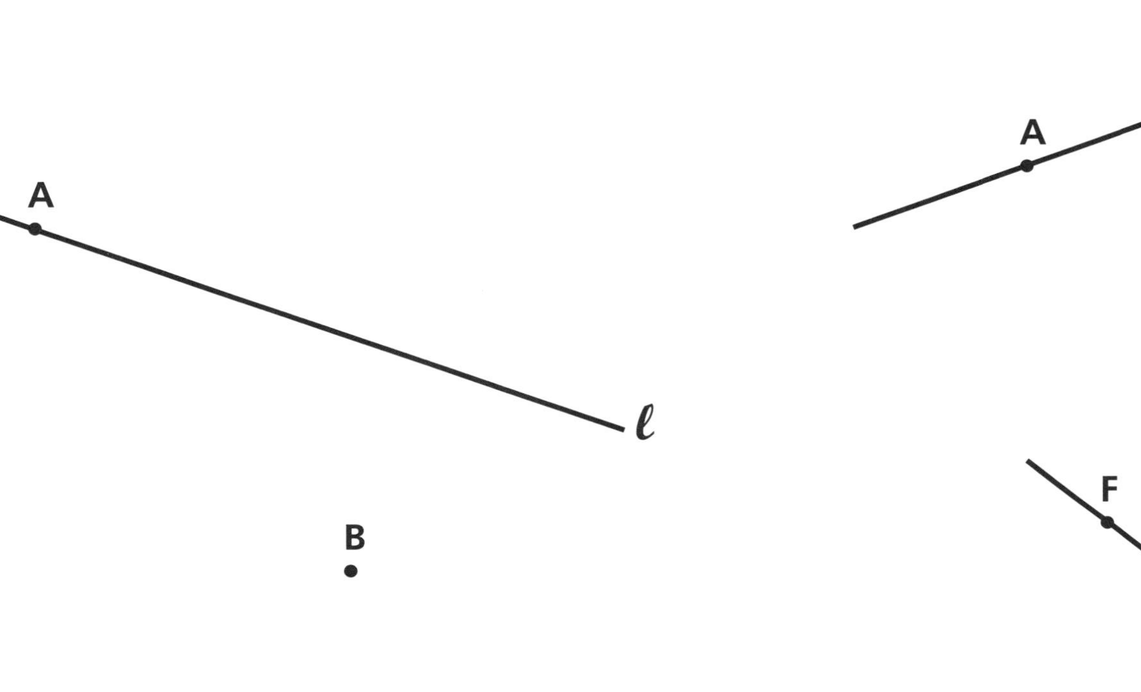

BONUS

Are the perpendiculars you drew parallel to each other? How do you know?

BONUS

How do you know that the angles of the rectangle and the square are right angles?

High Altitudes

Draw a line perpendicular to side $\overline{BC}$ that passes through A. Then draw a perpendicular through B to $\overline{AC}$ and through C to $\overline{AB}$.

TRIANGLE ALTITUDE

90°

An **altitude** of a triangle is the perpendicular from the vertex to the opposite side.

Draw the altitudes for each of the triangles below.

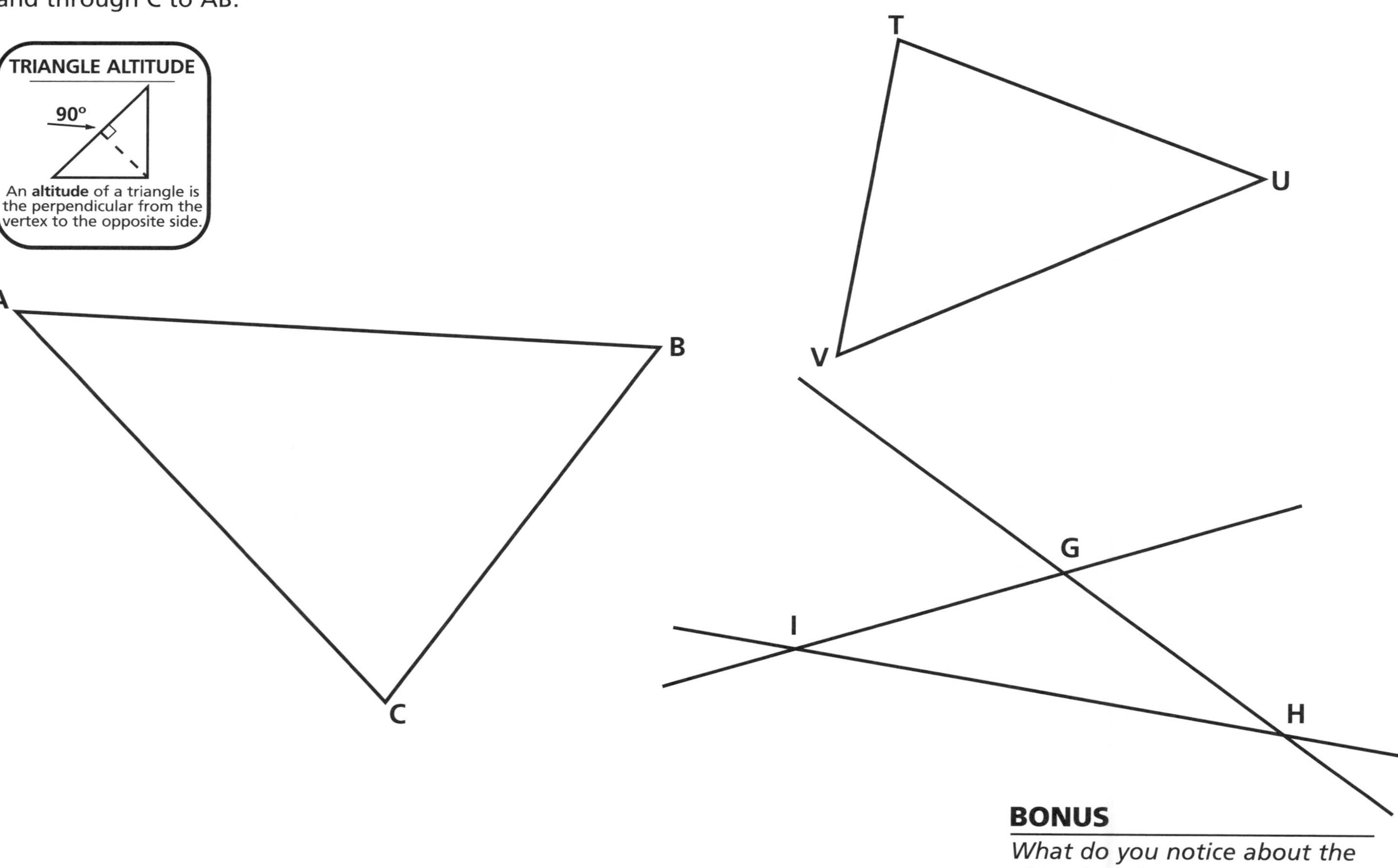

BONUS

What do you notice about the altitudes of any one triangle?

Mid-Point Crisis

Place the *GeoReflector* on AB so that the image of AB maps onto itself exactly. Mark the point, X, at which the *GeoReflector* crosses AB. Draw a line from X to vertex C of the triangle.

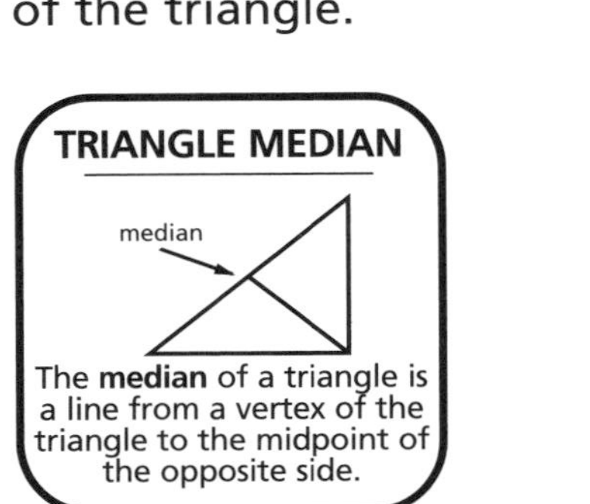

Draw the median for each side of each of the triangles.

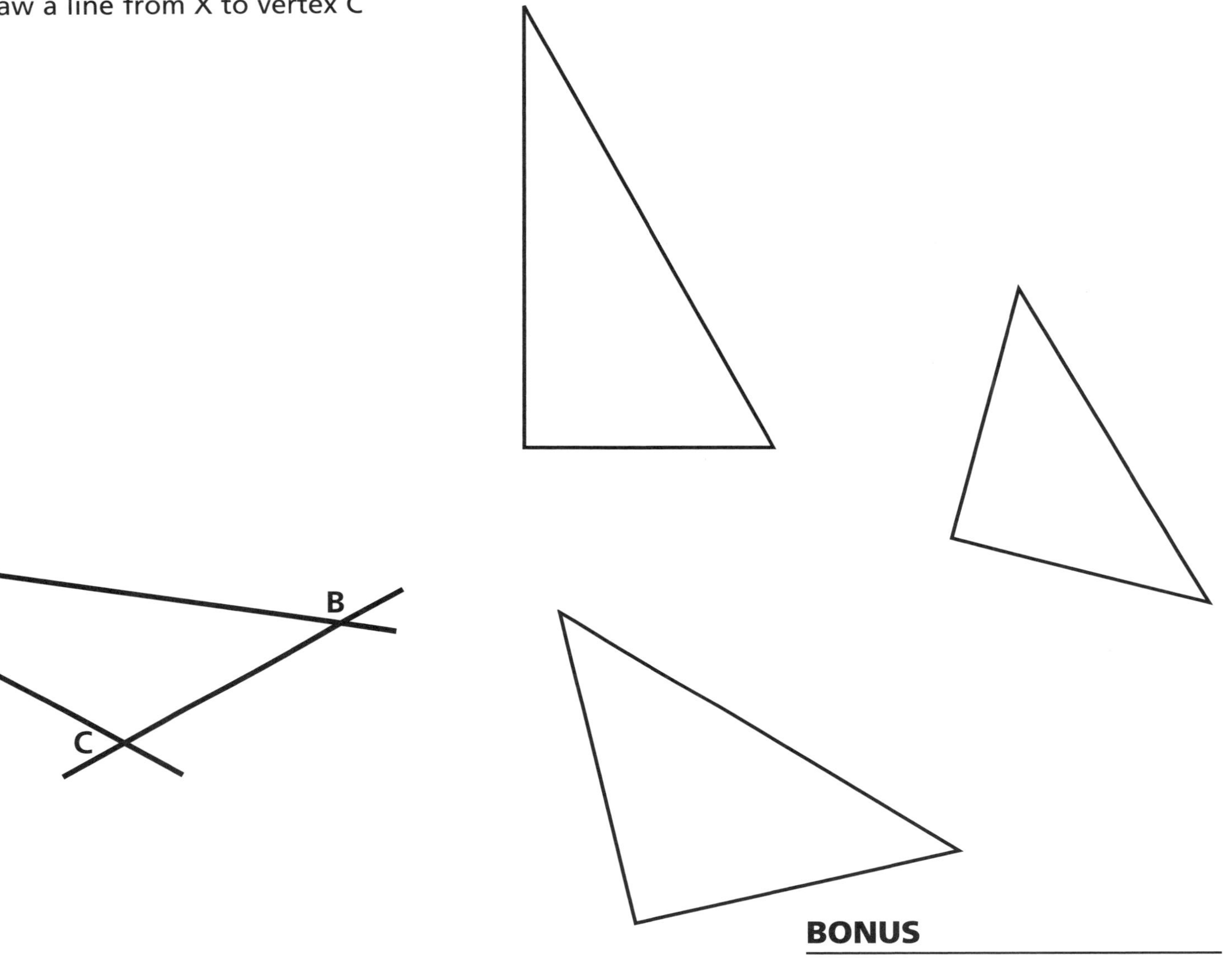

BONUS

What do you observe about the medians of these triangles?

Half a Line

Place the *GeoReflector* on ℓ so that the image maps onto itself. Slide the *GeoReflector* along ℓ until the length of the object line exactly matches the length of the image. Draw the **perpendicular bisector**.

PERPENDICULAR BISECTOR

ℓ

Construct a perpendicular bisector to ℓ_1. Then construct two more bisectors to make a rectangle.
HINT: Many rectangles are possible.

Bisecting Sides

Draw a perpendicular bisector to side BC of $\triangle ABC$. Then draw a perpendicular bisector to side AC and to side AB. Do the same for each side of the other triangles.

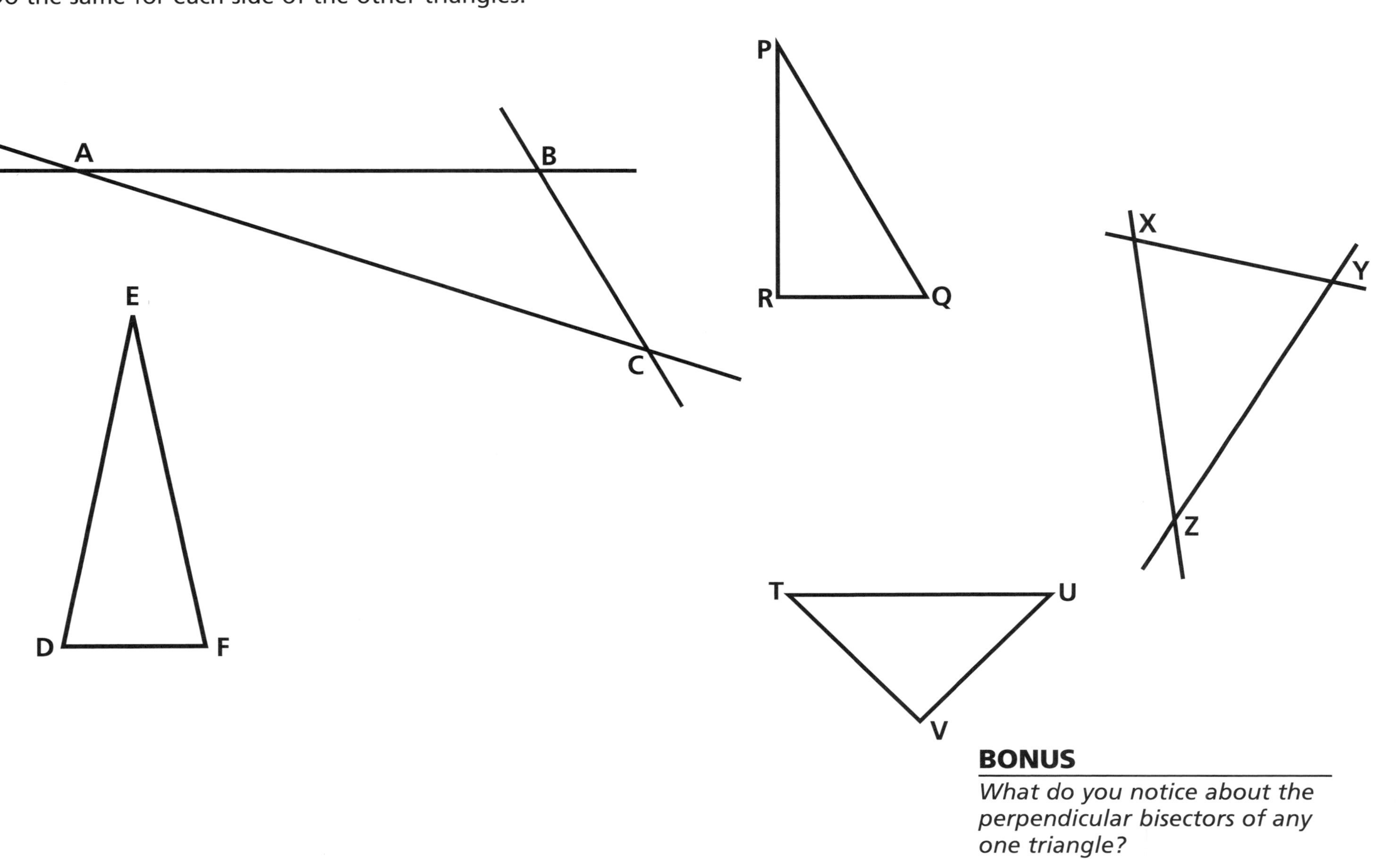

BONUS

What do you notice about the perpendicular bisectors of any one triangle?

Ring Around the Triangle

Draw a circle around ΔABC that passes through points A, B, and C. Do the same for ΔDEF and ΔTUV.
HINT: Find the point where the perpendicular bisectors of the triangle's sides meet. This is the center of the circle.

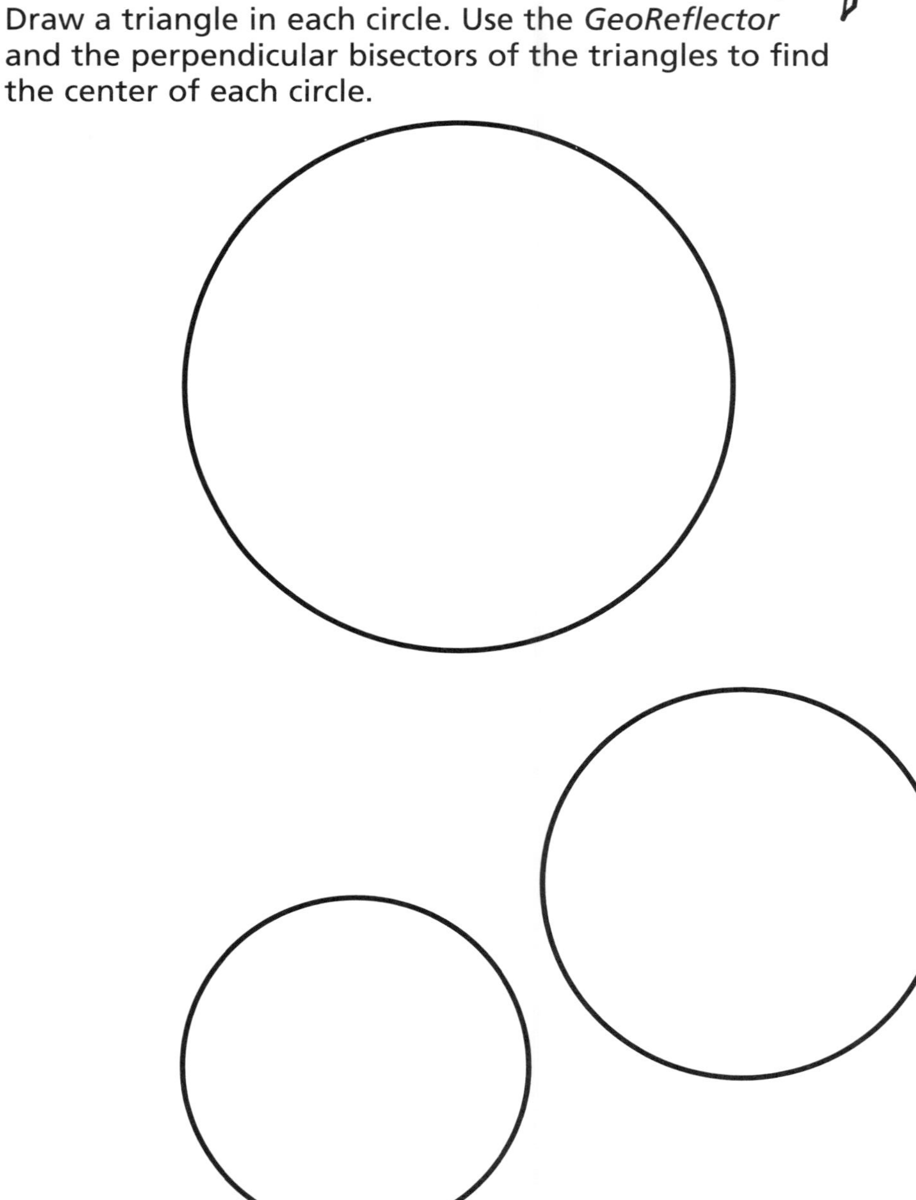

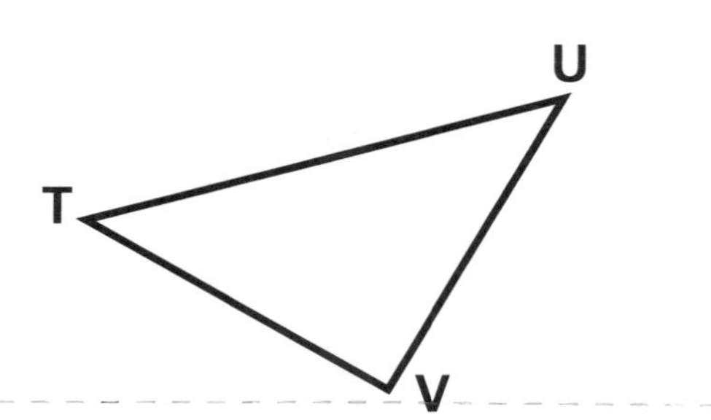

Draw a triangle in each circle. Use the *GeoReflector* and the perpendicular bisectors of the triangles to find the center of each circle.

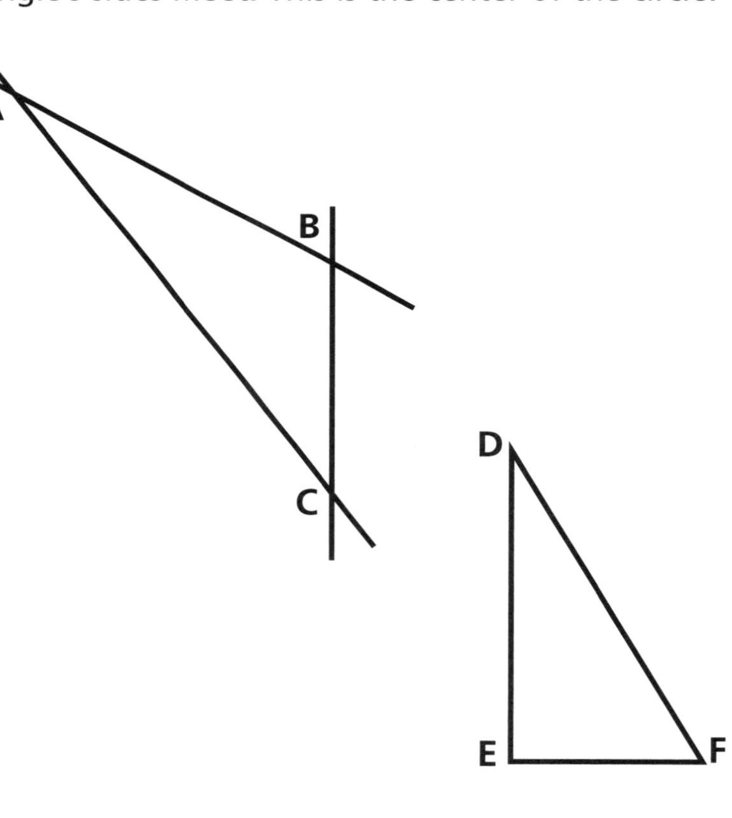

From Points to Circles

1. Choose a point X along AB of ΔABC. Choose a point Y along BC of ΔABC and choose a point Z along AC.
2. Find the perpendicular bisectors of the sides of ΔZAX, ΔXBY, and ΔYCZ.
3. Draw a circle around ΔZAX and a circle around each of the other two triangles, ΔXBY and ΔYCZ. Be sure the circles pass through the vertices of the triangle.

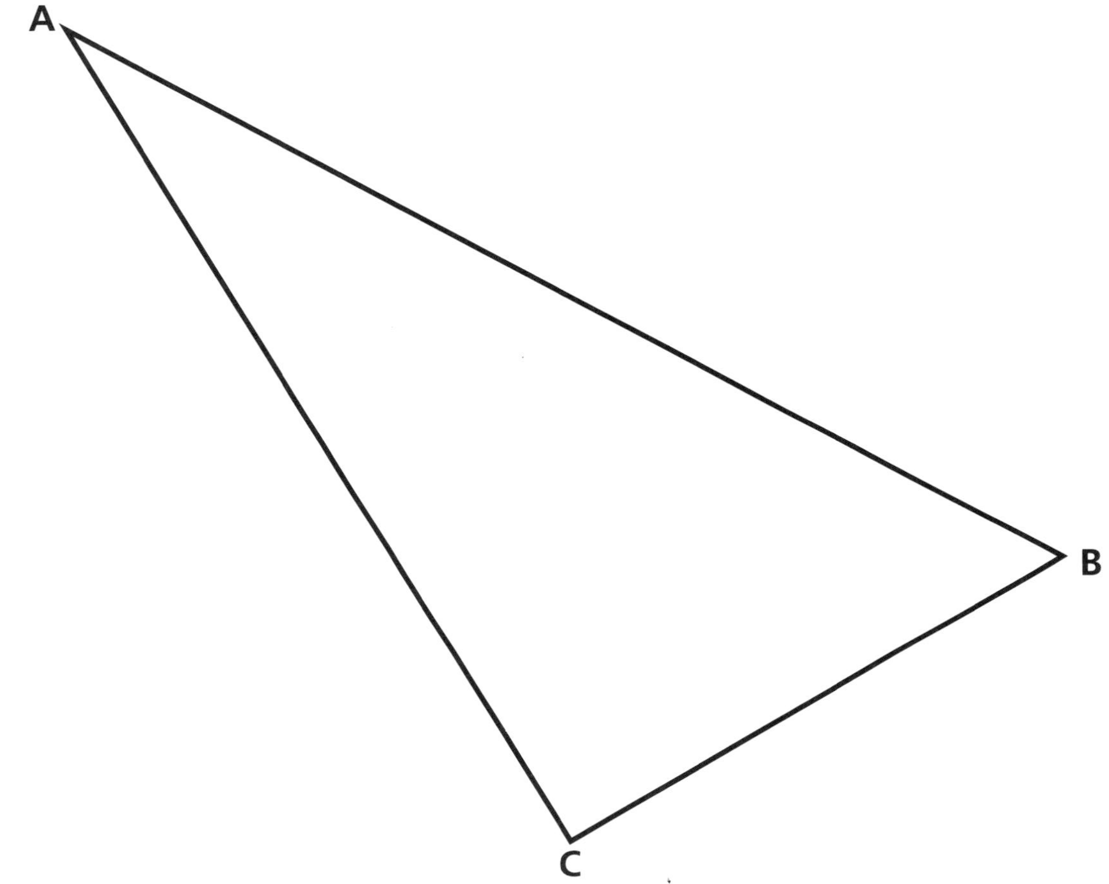

BONUS

What do you notice about the three circles?

Two Angles from One

Use your *GeoReflector* to map the image of $\overline{OA}$ onto line $\overline{OB}$. Draw a line to show where you placed the *GeoReflector* and place point C on the line. What can you say about the angle made by $\overline{OA}$ and $\overline{OC}$ and the angle made by $\overline{OB}$ and $\overline{OC}$?

ANGLE BISECTORS

An **angle bisector** divides an angle into two congruent parts.

Use the *GeoReflector* to bisect each of the angles below.

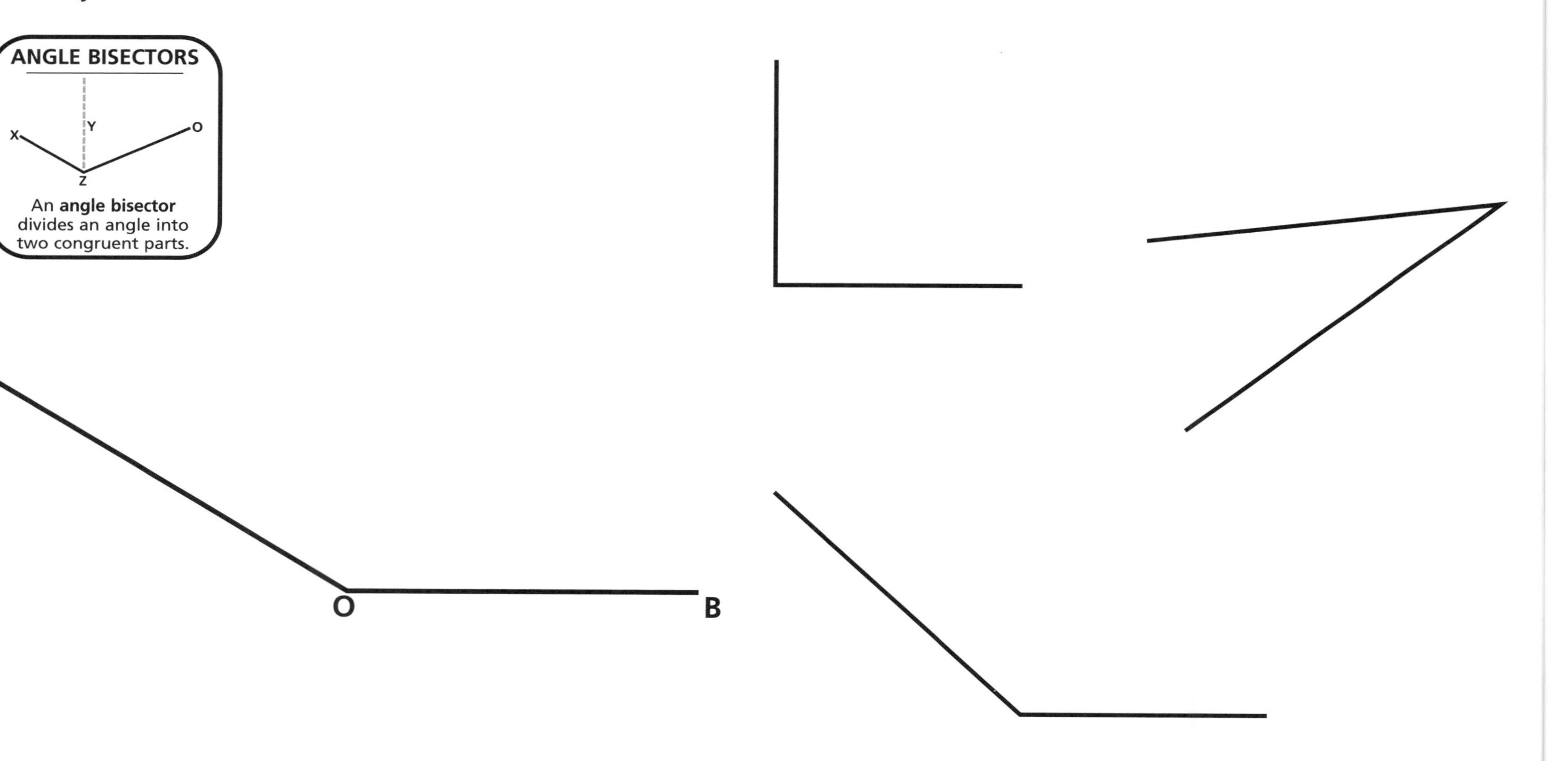

BONUS

How does the angle bisector compare to the line of symmetry for the angles?

Bisecting for Fun

Find and draw the angle bisector for each angle in the figures.

Find the angle bisectors for each angle in pentagon ABCDE. Find and draw the perpendicular bisector for each side of the pentagon. Do the same for equilateral triangle ZYX.

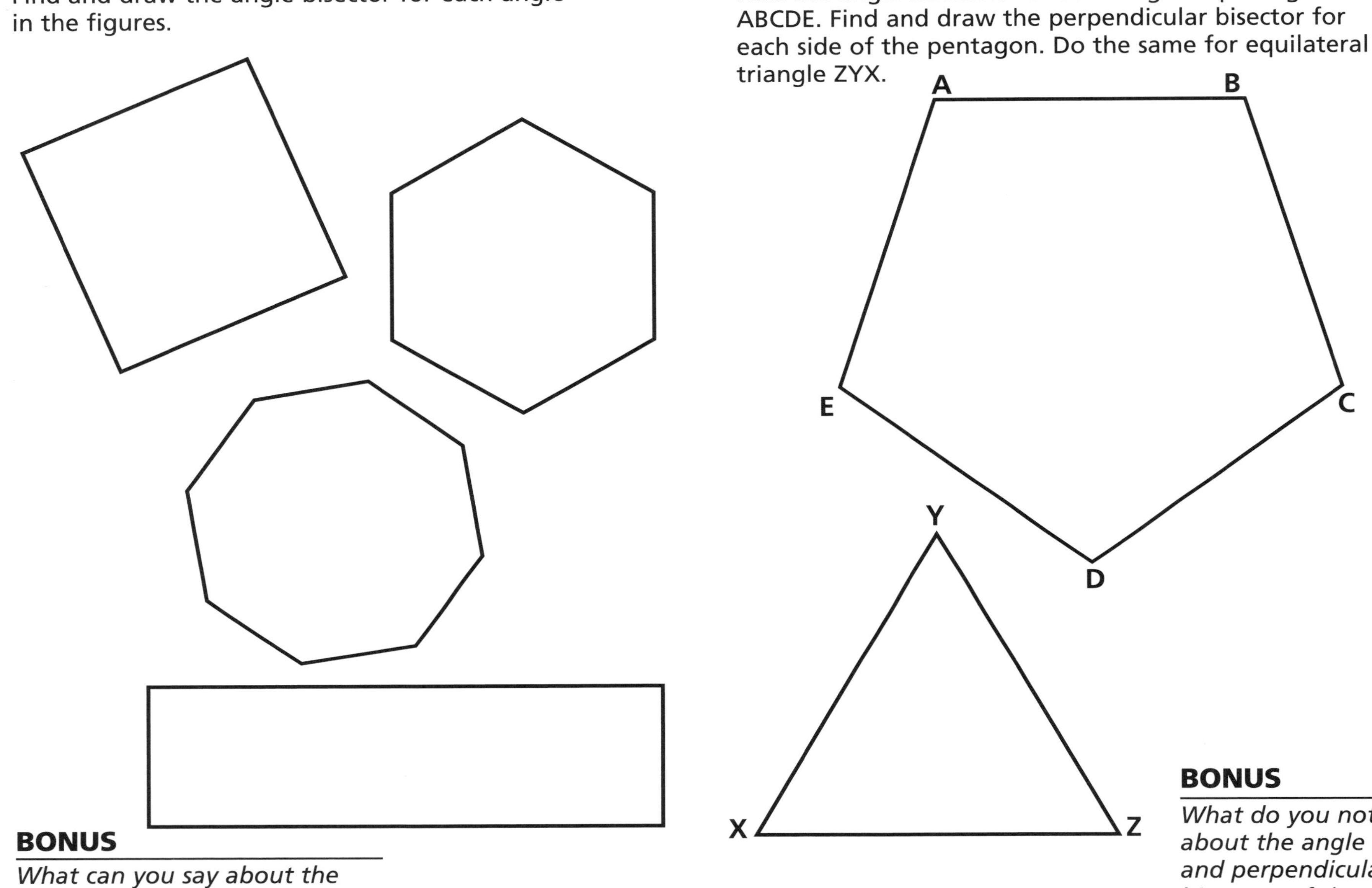

BONUS

What can you say about the angle bisectors and the lines of symmetry for each figure?

BONUS

What do you notice about the angle and perpendicular bisectors of the figures?

Inside Circles

Draw a circle that is completely inside ΔABC and touches each side. Do the same for ΔDEF and ΔHJK. HINT: Find the point where the triangle's angle bisectors meet. This is the center of the circle.

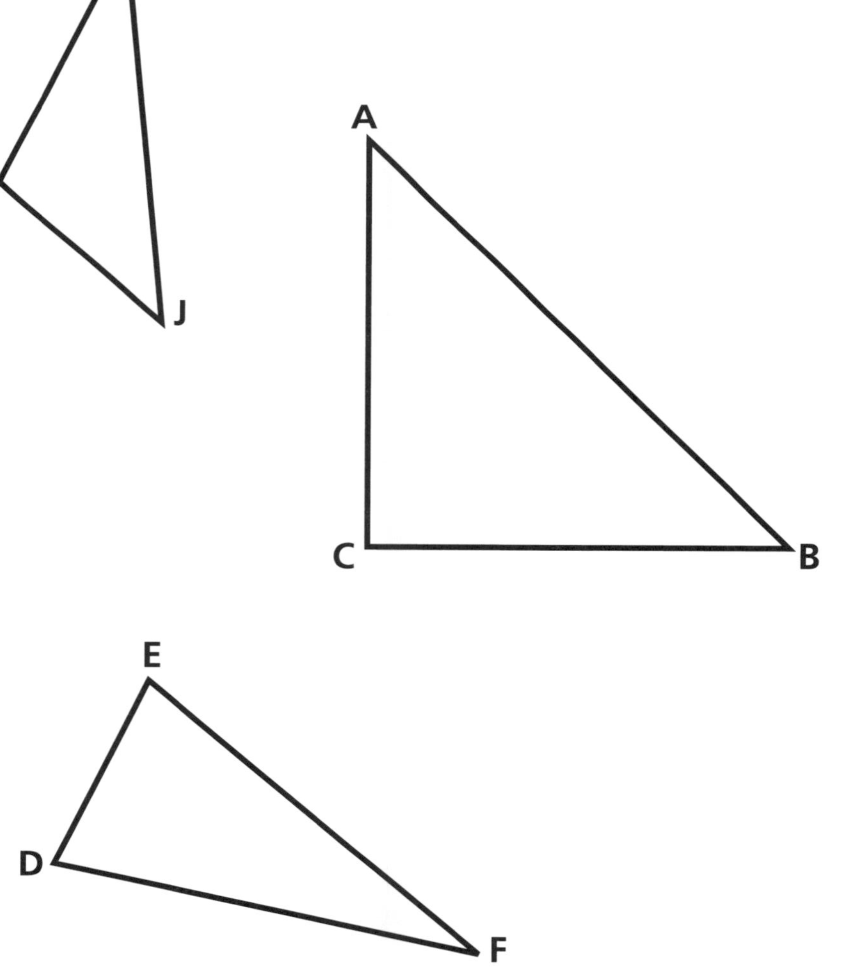

Draw a circle inside each triangle.

Circle Point Construction Zone

Using the points below, draw four different circles. Be sure each circle passes through three of the points.

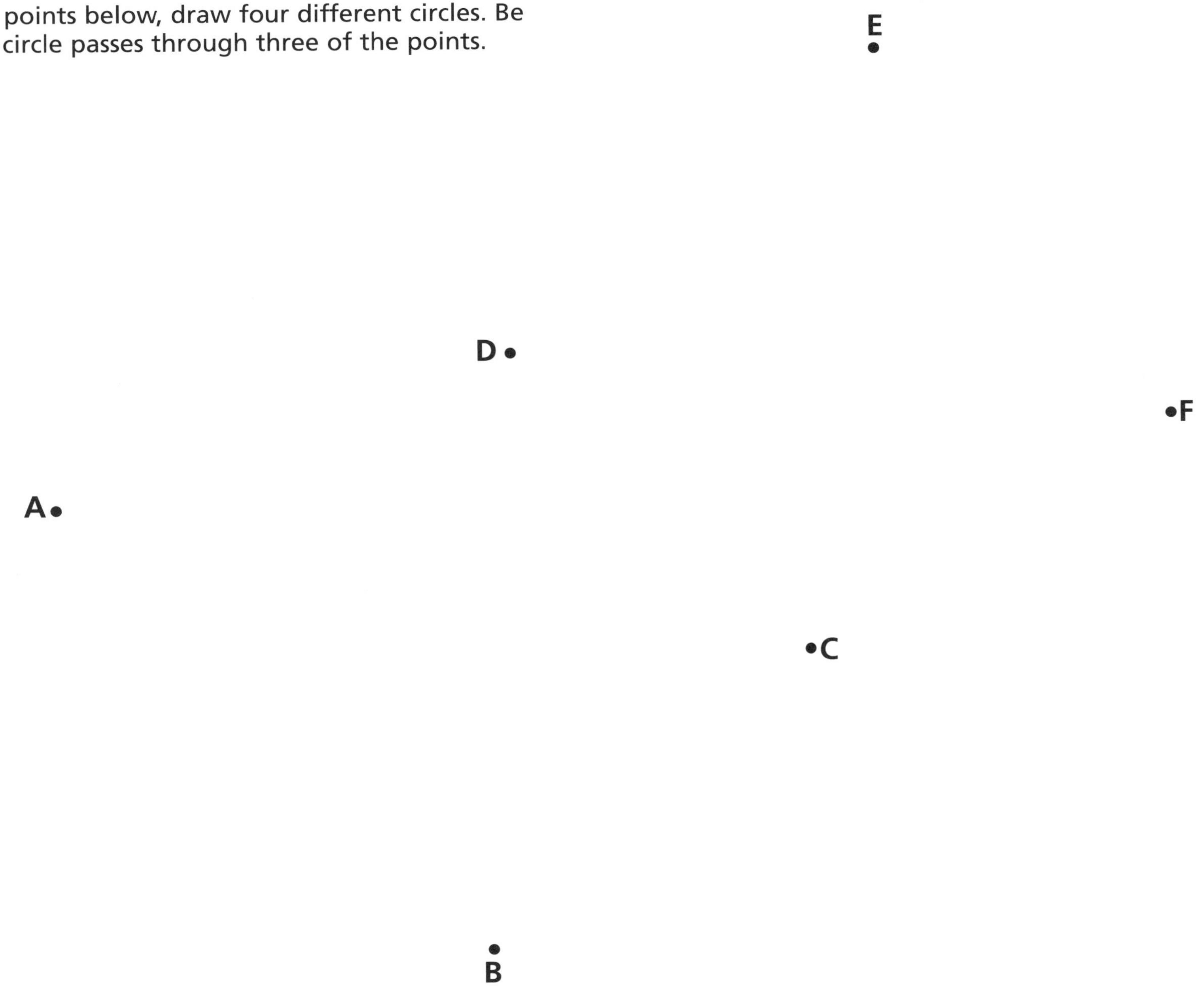

E

D

F

A

C

B

Square Deal Construction Zone

Use your *GeoReflector* and what you know about altitudes and perpendicular bisectors to draw a square in each of the three circles below.

Draw a **regular octagon,** a figure with eight equal sides, inside the circle.

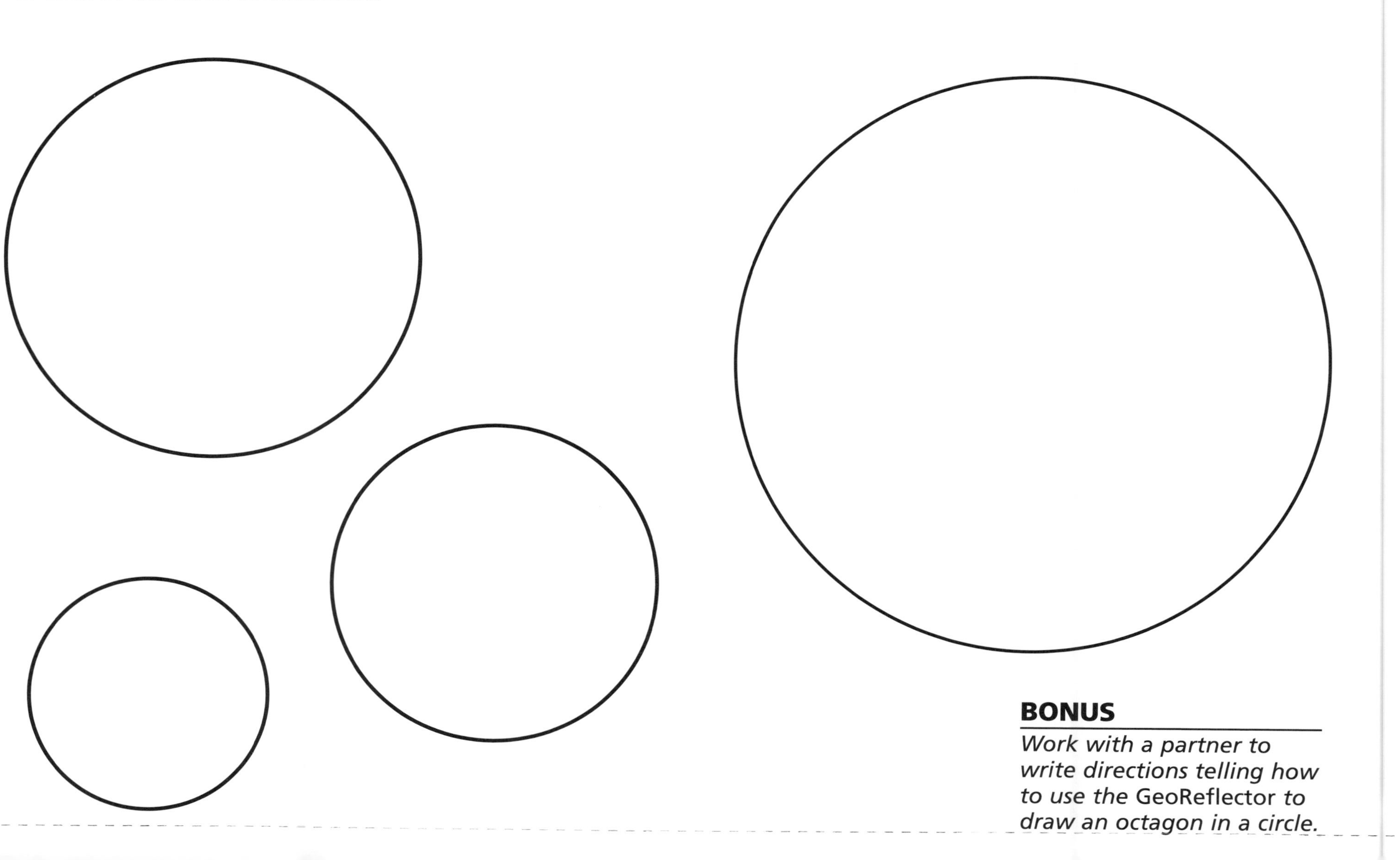

BONUS

Work with a partner to write directions telling how to use the GeoReflector *to draw an octagon in a circle.*

Triangle Challenge

Draw an **isosceles triangle,** a triangle with two sides of equal length. Use $\overline{AB}$ as one side of the triangle.

Draw an **equilateral triangle,** a triangle with all three sides the same length. Use $\overline{XY}$ as one side of the triangle.

BONUS

Work with a partner to write directions on how to meet the Triangle Challenge.

Circle This

Draw a triangle so that the center point of the circle which passes through the triangle's vertices is also the center point of a circle that touches each side. Use $\overline{AB}$ as one of the triangle's sides.

A B ℓ

BONUS

Describe the kind of triangle you used, and why it works.

Page Answer

7. The square is next to the image of the button, behind the *GeoReflector.*
8. The image moves whenever the object moves.
9. The size and shape of the image is the same as the size and shape of the object.
10. The two male gymnasts do handstands on the pommel horse.
11. Place the *GeoReflector* so it makes an image of two of the squares. The two image squares plus the 6 object squares make 8 squares.
14. ←— ←— is not possible because in the reflection, the arrowhead would point in the opposite direction.
18. The first reflection moves the eye patch to the right eye, a second reflection moves it back to the left.
21. The image moves twice as far as the *GeoReflector.*
23. The second slide image is the same shape and size as the first image, and the dots are in the same position. The rotation image is the same size and shape but the dots are stacked rather than next to one another.
25. The flipped tambourine player looks like the object player but the left and right sides change places.
26. A, H, I, M, O, T, U, V, W, X, and Y flip side to side. B, C, D, E, H, I, K, O, and X flip up and down. H, I, O, and X look the same after sideways or upside-down flips.
27. Yes. Flipped images have right and left or top and bottom reversed.
32. The symbol ☯ is not symmetric.
34. The line of symmetry is the same distance from each item in the pair.
35. The parallelogram does not have lines of symmetry. A circle has an infinite number of lines of symmetry.
37. Corresponding points are the same distance from the line of symmetry.
38. If the image of one object exactly maps onto another object, the two items are congruent.
39. Turn the tracing paper over to see if it maps exactly onto the second snack food.
40. Use reflections and tracing paper to map one triangle onto another.
45. See which lines map exactly onto one another.
46. There are an infinite number of perpendicular lines.
47. Yes. They pass the reflection test because they were drawn using perpendicular lines.
48. The altitudes meet at one point.
49. The medians meet at one point.
51. The perpendicular bisectors meet at one point.
53. They intersect at one point.
54. It's the same.
55. The angle bisectors and lines of symmetry for squares, octagons, and hexagons are the same. The angle bisectors and perpendicular bisectors of a regular pentagon and an equilateral triangle are the same.
58. 1) Draw a triangle in the circle. 2) Find the center of the circle by finding where the $\perp$ bisectors meet. 3) Draw one diameter, then find its $\perp$ bisector. 4) Connect the endpoints of the diameters to make a square. 5) Find where $\perp$ bisectors of the sides of the square meet the circle. 6) Connect the points of intersection along the circle.
59. To draw the isosceles triangle, draw $\overline{AB'}$, the reflection of $\overline{AB}$. Connect B′ and B. To draw the equilateral triangle, find the $\perp$ bisector of side $\overline{XY}$. In an equilateral triangle, this is also the angle bisector of the opposite angle. Place $\overline{XY'}$, the reflection of $\overline{XY}$, so that it meets the $\perp$ bisector. Draw $\overline{XY'}$, then $\overline{Y'Y}$.
60. An equilateral triangle works because the angle bisectors and the $\perp$ bisectors meet at the same point.

Selected Solutions

PAGE 48—High Altitudes

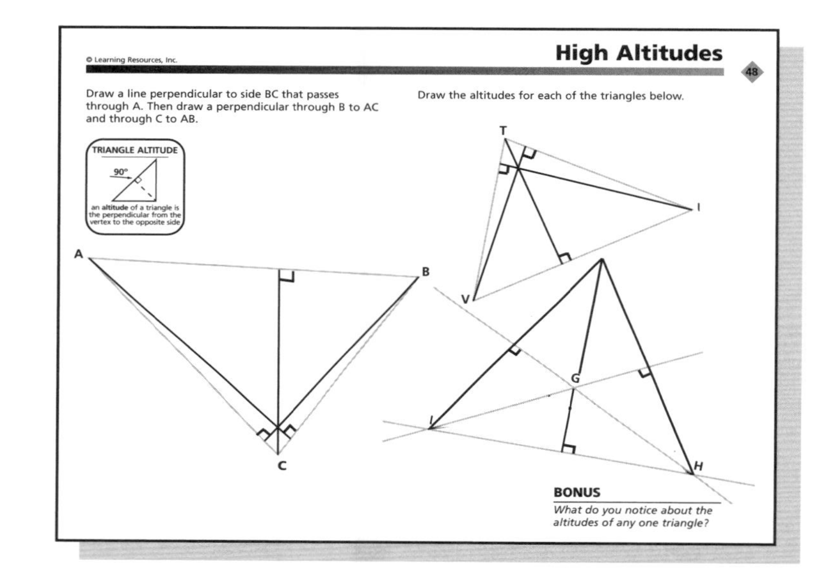

PAGE 49—Mid-Point Crisis

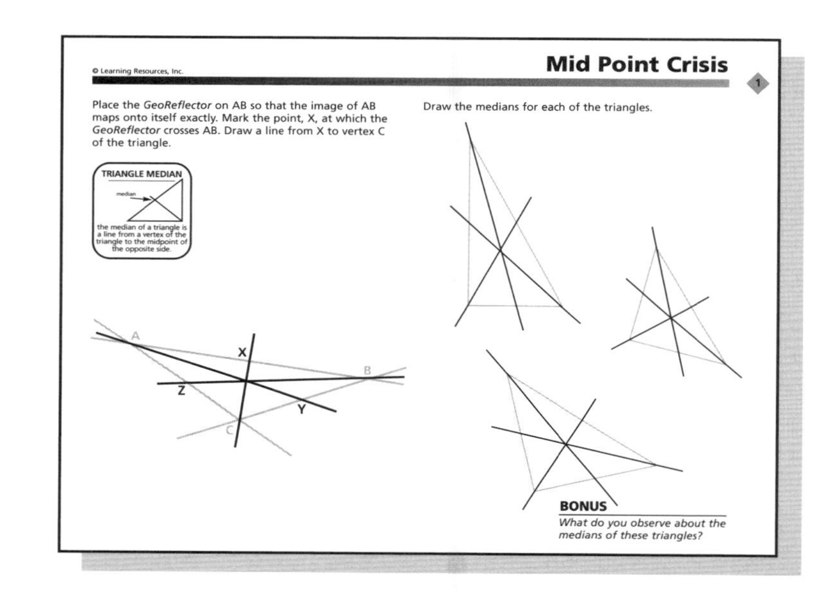

PAGE 50—Half a Line
(students answers may vary)

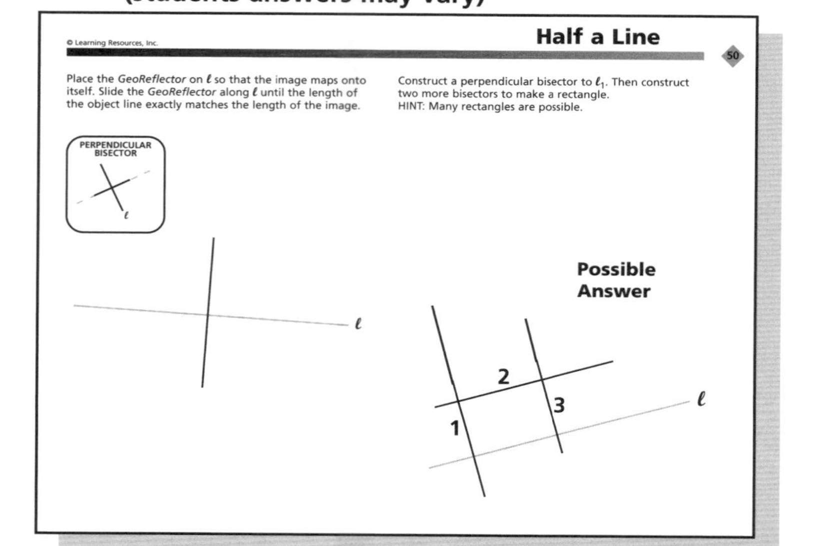

PAGE 51—Bisecting Sides

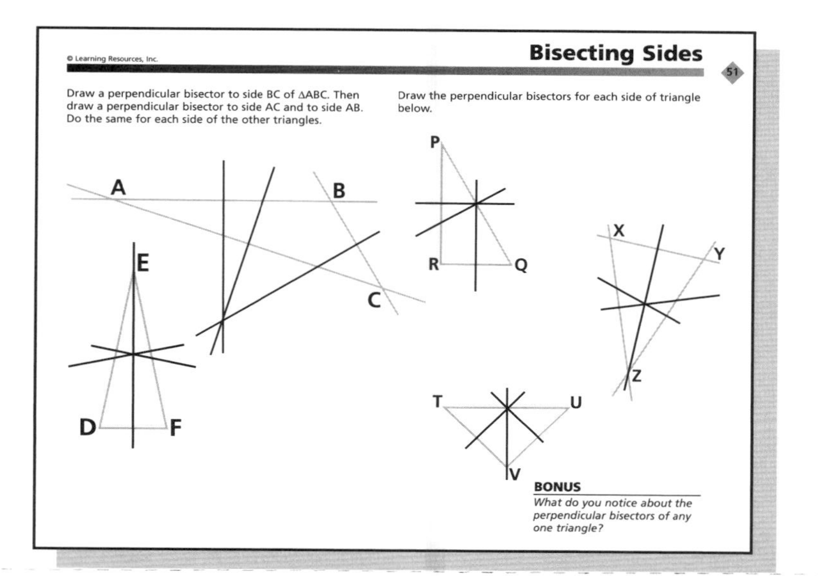

PAGE 52—Ring Around the Triangle
(students answers may vary)

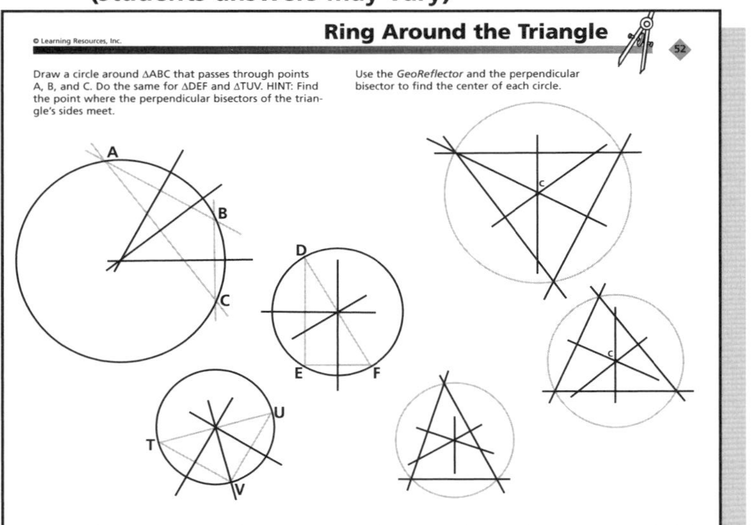

PAGE 53—From Points to Circles

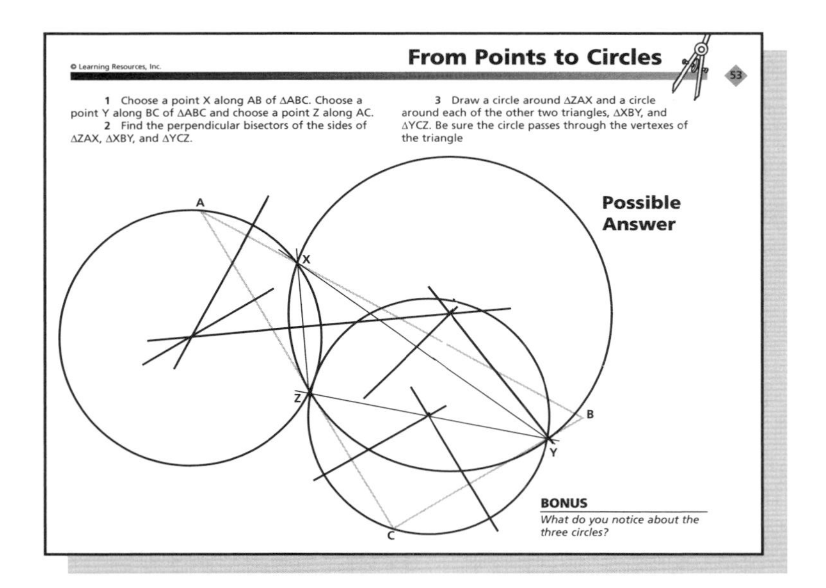

PAGE 55—Bisecting for Fun

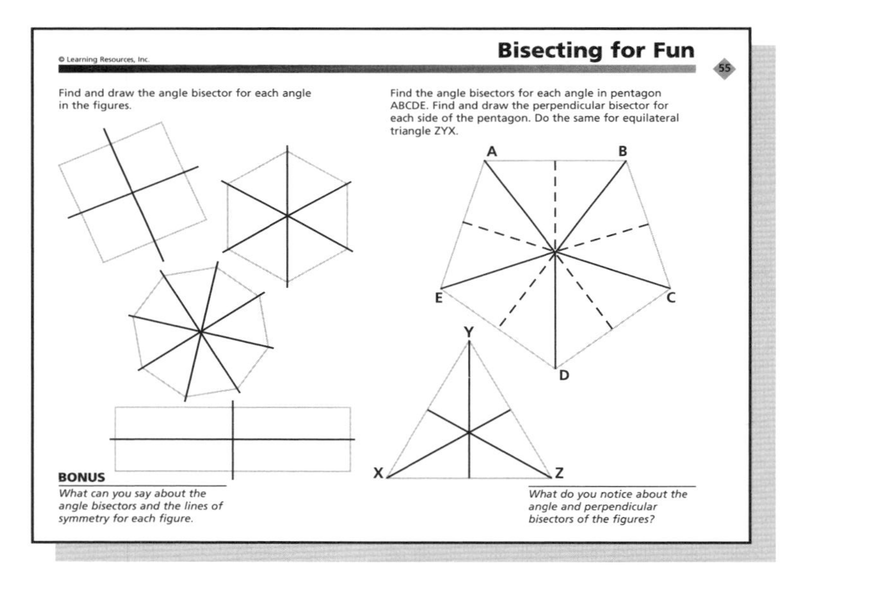

PAGE 56—Inside Circles

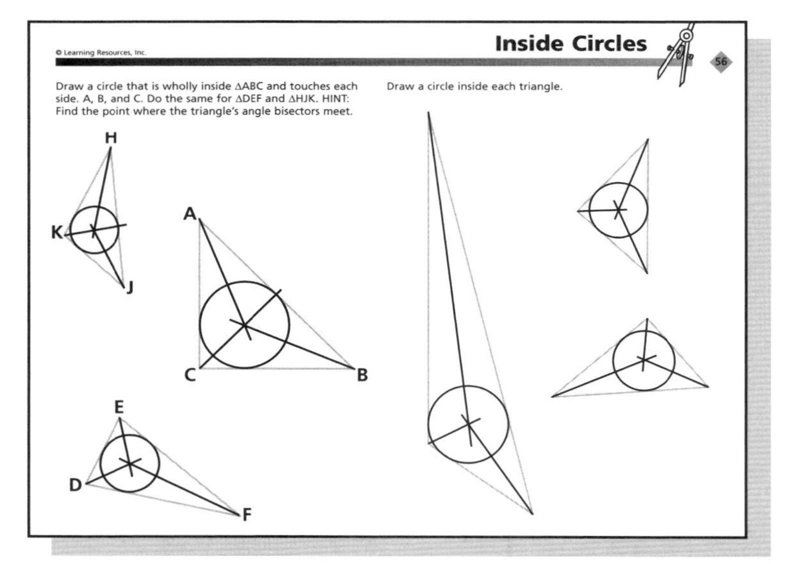

PAGE 57—Circle Point Construction Zone

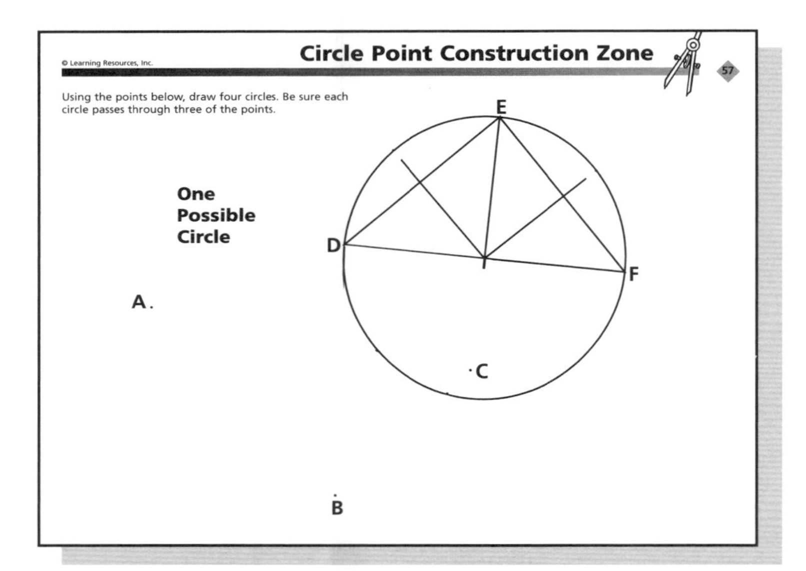

PAGE 58—Square Deal Construction Zone

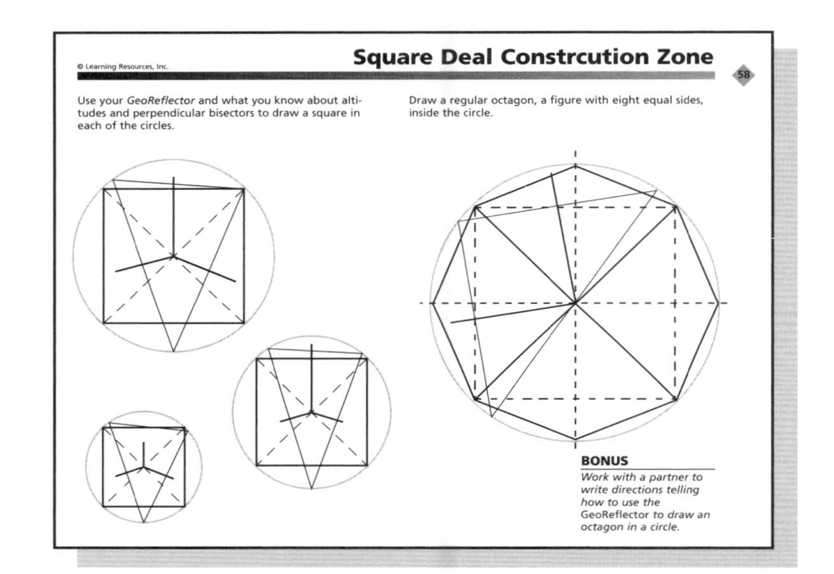

PAGE 59—Triangle Challenge

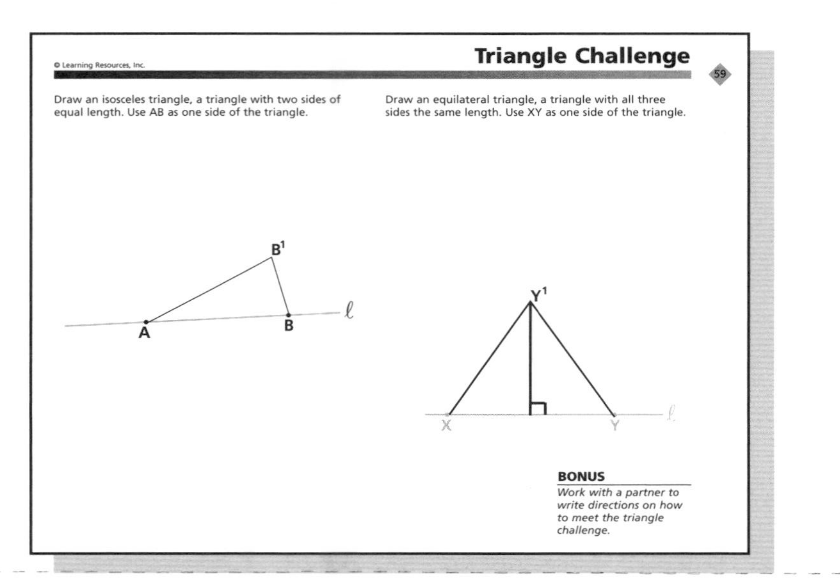

PAGE 60—Circle This

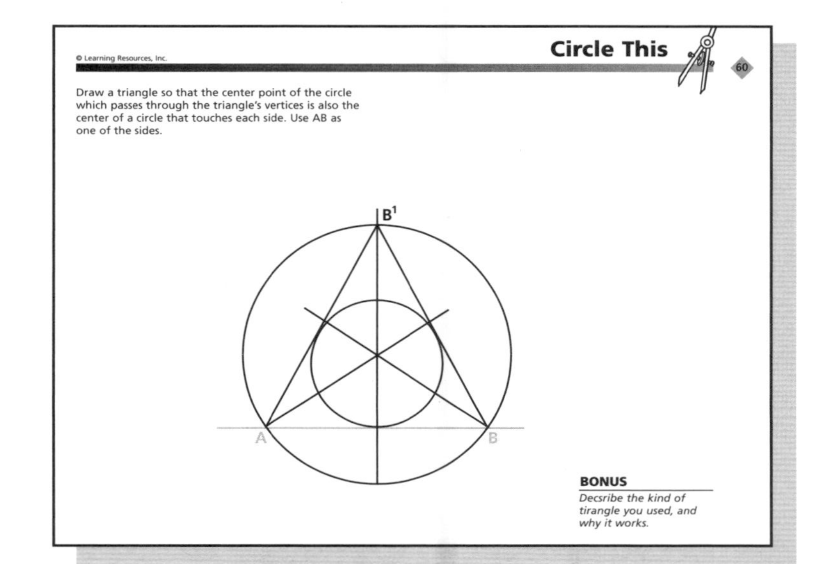